The Fast Track to Understanding Ham Radio Propagation

I0477596

by Michael Burnette, AF7KB

Dedication

John Van Dalen, N7AME

Great friend, great ham, great community servant,

and lifelong learner.

Very special thanks to my physicist friend, Dave Cornell, PhD, W9LD for his always gentlemanly guidance of the science in this book.

Never be afraid to try something new. Remember: amateurs built the ark; professionals built the Titanic.

- Anon

Contents

Introduction

Propagation is "everything that happens to our signal after it leaves our antenna." It's the process by which our electromagnetic waves are conveyed from one location to another.

Propagation begins when we send a current through our transmitting antenna. Instantly, an electromagnetic field starts radiating through space at something closely approaching the speed of light. If we're transmitting at 7.180 MHz, there are 14,360,000 fields flying out of our antenna every second.

Each of those fields has several dimensions;

- Amplitude; the strength of the field.

- Direction.

- Wavelength.

- Polarization; vertical, horizontal, circular, or elliptical.

- Velocity; how fast it is moving.

If we were living in intergalactic space, sending our QSO's through a vacuum far, far away from any terrestrial or celestial objects, this book on propagation would be a great deal shorter. Deep space is what we might say is an "ideal" medium for propagation. Physicists would say it is, "A linear, isotropic, nondispersive, homogeneous and non-time-dependent medium."

- Linear; conducts the wave the same throughout the path.

- Isotropic; no directionality

- Nondispersive; doesn't affect polarization of the wave.

- Homogeneous and non-time-dependent; has the same dielectric properties throughout the path all the time.

In that environment, we'd just crank up an appropriate amount of transmitter power, perhaps point our highly directional antenna at our target, and blast away. Essentially, what we sent out would be what was received, every time. Propagation problems solved!

Down here on Earth, things are a lot more complicated. We don't have *any* linear, isotropic, nondispersive, homogeneous and non-time dependent signal paths available to us on any frequency. The Earth itself forms a colossal barrier to transmission of most of our signals, the ground (and even the sea!) is uneven, there are countless man-made objects around with wildly varying effects on our signals, the fickle ionosphere is constantly changing how it treats our signals, and even the very air we breathe can affect our transmissions.

Here on Earth, our blobs of electric field fly out of our antenna and immediately start interacting with all sorts of matter; and, every encounter with matter affects each field in some way, altering at least one and maybe all but one of those dimensions. (So far as we know, charge remains unchanged.)

Ever since Marconi sent his famous signal across the Atlantic in 1901 – defying detractors who said it was impossible because radio waves could only travel on line-of-sight paths – we've known there was more to propagation than met the eye.

For just as long, we've also known that there was tremendous commercial potential and military advantage to be gained through better understanding of propagation mechanisms, so enormous amounts of research have been and continue to be devoted to the science. Ham radio has both contributed to that research and benefited greatly from it.

Despite all that research there are still many unknowns and unpredictables in the science and precious few "magic bullet" solutions. That's especially true for us ham radio operators because of the limited range of frequencies and amount of power available to us, but there certainly are "best practices" that will raise our chances of making the contacts we seek. Not all of them involve erecting a 199 foot tower in the back yard, either, though that never hurts!

The science of propagation is filled with elegant and weighty mathematical formulas such as the ever popular:

$$F_{L_s}(I_s) = \mathcal{P}(L_s \leq I_s) = \int_{\infty}^{I_s} dl\, f_{L_s}(l)$$

We, however, are going to proceed forward with a minimum of high-falutin' math and a maximum of plain language explanations when we do use some math. If you're interested in the formulas, you'll want a slim (but densely packed!) book, *Radiowave Propagation*, by Curt A Levis, Joel T. Johnson, and Fernando L. Teixeira. Warning: It's a textbook for Electrical Engineering grad students and seldom goes more than three sentences without a formula that looks like the above.

As a ham you really don't need to be able to do all the math because there are remarkably powerful propagation prediction software programs available to us – at least one of which is absolutely free.

It's not my intention here to turn you into an Electrical Engineer, so I've simplified many concepts. I want to give you what's called a "working knowledge" of propagation – it won't be a high level academic understanding.

No one enjoys for very long the experience of sitting in front of a radio and failing to communicate, so I hope I can point you toward at least a few tricks you might have missed along the way, and maybe open up some areas of amateur radio you haven't tried yet. I hope, too, to enhance your enjoyment of the hobby as you come to understand how signals travel between you and that other ham.

I think hams have the most curiosity about ionospheric propagation, which is good, because it is definitely the propagation type with the most to learn, and it's the one we'll spend the most time on in this book. Other modes are useful to us, though, and we'll cover those as well.

Propagation is a function of many factors. Some are completely under our control, some are partially under our control. Some are dictated by the rules and regulations of the amateur service, and some are just more or less happenstance. Propagation can be fickle, and nothing you read here is graven in stone. It's not unusual for Super Ham, with all the gear and all the watts, following all the best practices, to come up empty-handed while the joker down the street with an old swap meet TS-520 tunes up his barbwire fence and scores DX all night.

One essential element in any propagation problem is frequency. Every form of propagation is dependent on the frequency of our transmission. While they're both electromagnetic waves, a 160 meter wave and a 70 cm wave behave wildly differently after they leave our antennas. In some conditions, even a difference in frequency of a kHz can make the difference between making a contact and listening to static.

Another essential element is the specific nature of the medium through which we plan to send our signal. There's no such thing as a generic medium. For ground waves, for instance, our signal's propagation will be affected by the conductivity of the particular ground between us and the receiver. In the continental United States, ground conductivity can vary from area to area by a factor as large as 30x. Even the "straight-line-of-sight" path of a UHF signal can be bent, distorted, and scattered by variations in the air through which it travels.

From the ground up, then, the various types of propagation we will cover include:

- Ground wave propagation
- Earth-ionosphere waveguide propagation
- Line-of-sight propagation & Tropospheric scatter propagation
- Tropospheric ducting propagation
- Ionospheric propagation
- Meteor propagation
- Auroral propagation
- Space propagation

Chapter 1 -- Ground Wave Propagation

A ground wave is what Marconi thought he had somehow created when he sent that message to a ship at sea that was over the horizon. (At least we think that's what he thought. We really don't know a whole lot about how Marconi did what he did, nor what his thoughts were. He was a lot more interested in owning the patent for radio, attracting investors, and keeping his secrets than he was in science.) He thought he had made a wave that followed the curvature of the Earth. He was wrong, because he could not have done that with the antenna he used, but later we did learn how to do that.

If you could see a ground wave, it might look something like this:

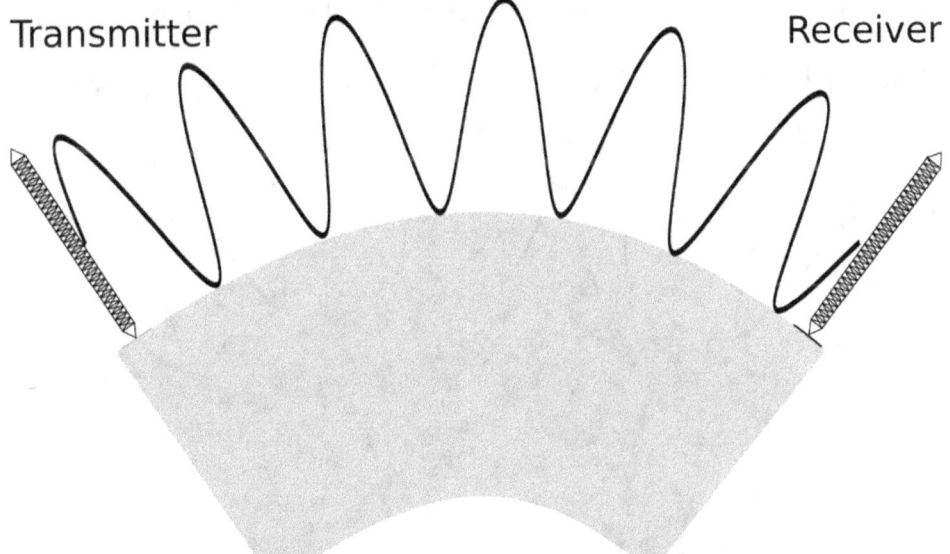

Ground waves are almost exclusively a low to medium frequency phenomenon. It's difficult to even create a ground wave at frequencies above 3 to 5 MHz, and even if you create it, it won't make it far. Even at ideal frequencies, a ground wave is, at best, a medium distance propagation tool. They're great for short distances; commercial AM stations, down there on frequencies just below our 160 meter band, spend large sums of money to create strong ground waves.

Within their limits, ground waves are superbly dependable. They never fade, they don't disappear if you get a building between you and the transmitter, and they don't stop working because we're at the bottom of a sun spot cycle.

They'll roll right over the horizon and keep going until they get so attenuated that they're gone.

They run out of steam for precisely the same reason that they hug the ground and can get over the horizon.

When a ground wave is created, it's really two waves. One above the ground and one down in the ground. At any given point, the wave below the ground is the opposite polarity of the one above the ground.

Because opposite charges attract, and because the wave in the ground is moving more slowly than the above ground wave, the bottom of the above-ground wave front slows down and the top (sort of) tilts forward, almost like an ocean swell breaking as it hits the friction of a shallow bottom.

Transmitter   Receiver

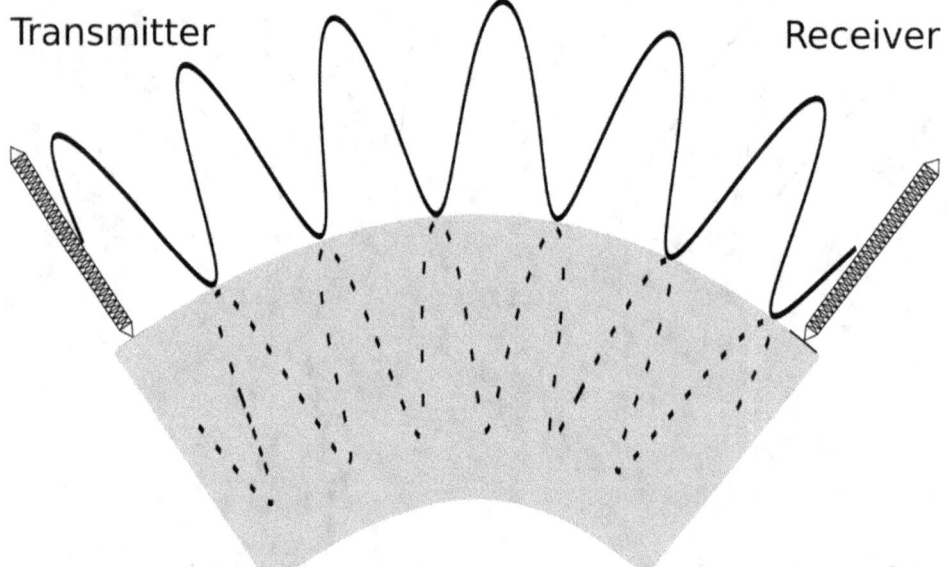

That mechanism that makes ground waves so useful also spells their doom. Not only is the Earth a lossy conductor of radio waves, it's also a slow conductor of those waves, so the below-ground wave is starting to get out of phase with the above-ground wave the instant they leave the transmitter. Chaos quickly ensues, and the ground wave has gone as far as it can go. That's why ground waves are useless at HF wavelengths – the two waves rapidly go out of phase with each other. (Earth, generally speaking, isn't very effective at conveying higher frequencies, anyway. If it did, we'd all have our HF antennas aimed at the ground.)

Practically speaking, for hams, effective ground waves only exist in the 160 meter band.

Pure ground wave communication, for us, probably tops out at a range of 50 miles – and even that would be quite extraordinary. Consider that a big commercial AM station, operating on a frequency far superior to ours for ground wave propagation and running a 50 kW transmitter into ¼ wave antennas with ground radials strung out a full ¼ wave every 3 degrees or so all the way around the tower struggles to cover a 100 mile radius, and you begin to appreciate what a challenge we're up against.

Still, for dependable crosstown communication, ground waves are a lot more dependable than, say, trying to use NVIS (Near Vertical Incident Skywave) off the ionosphere.

Before my mail box explodes with furious letters of protest from 160 meter fans, let me make it very clear that I am not saying 160 meters is only good for, at best, 50 mile communication. Far from it – nighttime 160 meter skywave DX can be spectacular.

There are three keys to creating a strong ground wave.

First, bring lots of watts. If your ham shack doesn't feel like a sauna, your amplifier needs more power.

Second, have an awesome ground system for your antenna. You'll probably want a vertical antenna or an inverted L (which is usually a lot more economical choice), but either one needs a really good and extensive set of ground radials either laying on the ground or buried a few inches down. (Pedantically speaking, if it's above ground, it's not a radial, it's a counterpoise.)

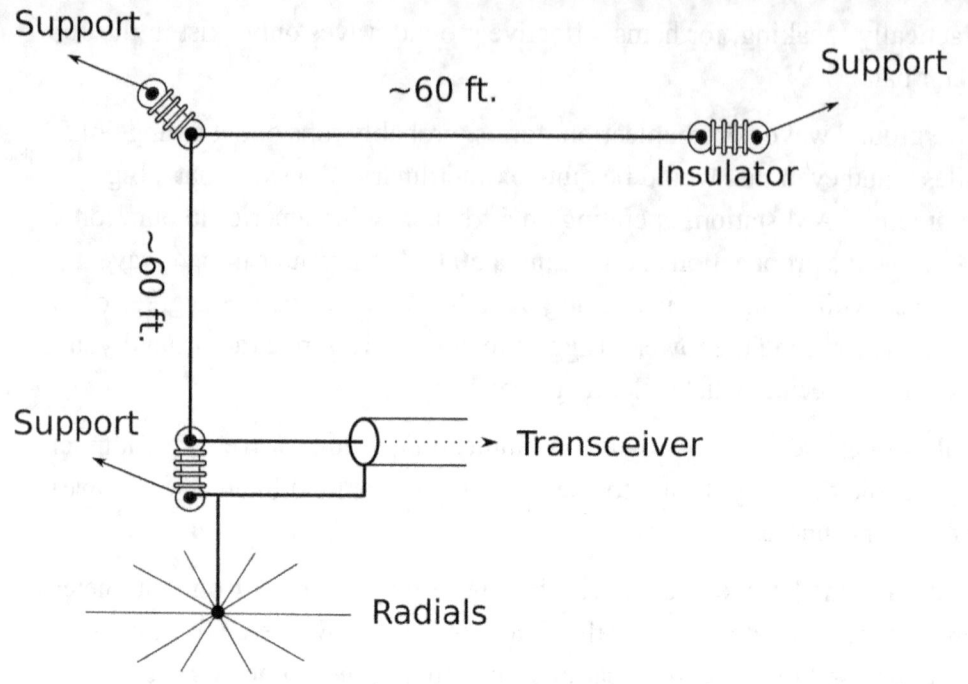

One version of an Inverted-L Antenna

Third, have the wisdom and foresight to live in an area of the country where the soil has excellent conductivity.

Even better, live by the sea and get your radials in touch with that salt water. Ground wave propagation over sea water can (at least theoretically) reach up to five times farther than ground wave on "very dry land", although as you look at the map on the next page of US ground conductivity, you'll see that there's some fine ground wave land that's pretty dry.

The map shows the conductivity in "millimhos per meter." Mhos ("moze") are the reciprocal of ohms, so higher numbers are better for ground waves. (Although this is an official FCC national ground conductivity map, their terminology is more than a little out of date. Today the term for mhos is "siemens.") The peak value is around 30 millimhos per meter, and if you look closely at the fine print, you'll see that seawater is "assumed to be 5000 millimhos per meter.

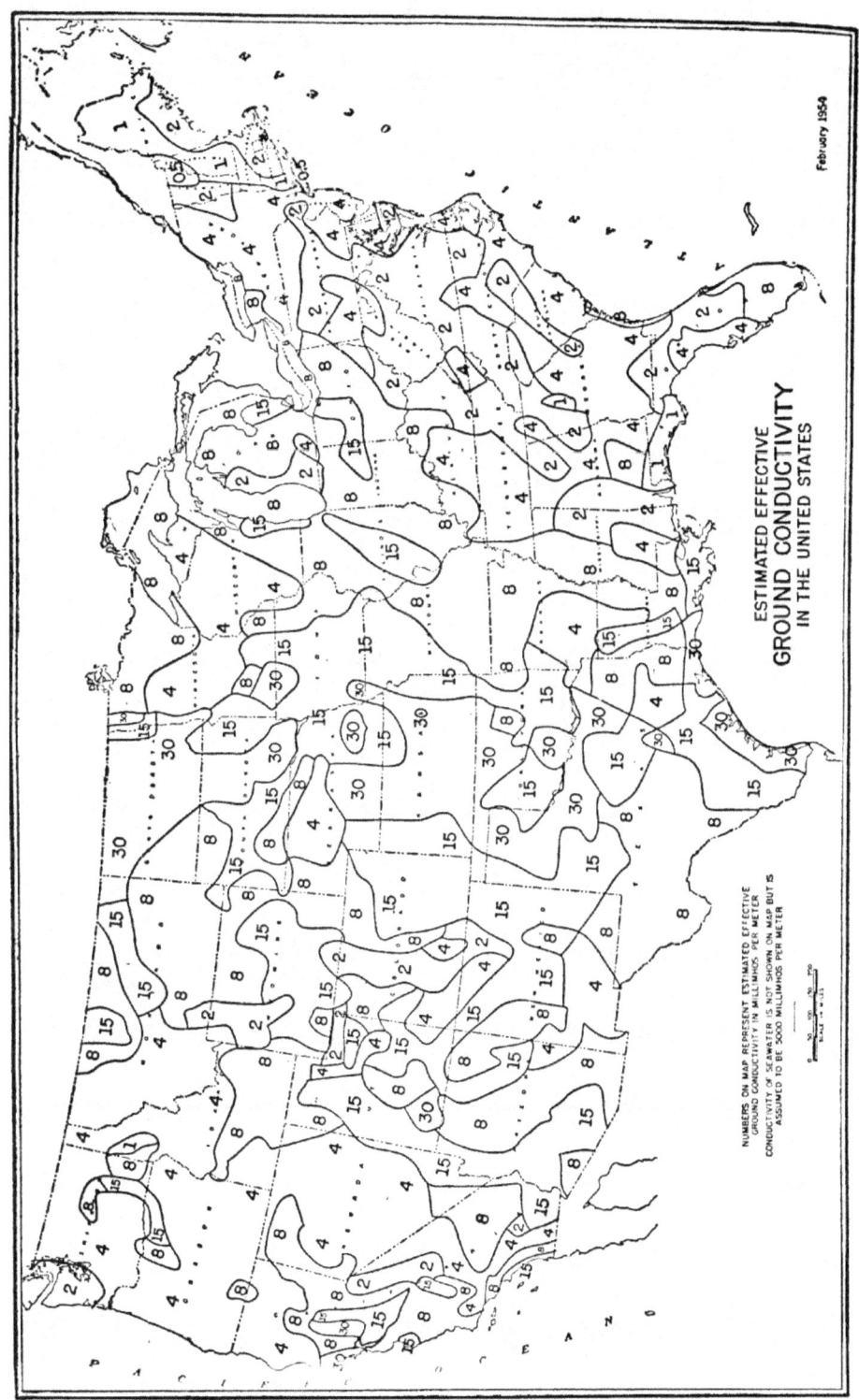

Ground Conductivity in the United States - millimhos per meter

17

Chapter 2 – Earth-Ionosphere Waveguide Propagation

Let me break the news about this one early and as gently as possible: You're probably not going to be using this one unless you're a really ambitious experimenter or just happen to be a communication specialist in the military. However, it's an interesting one to know about.

Let's start with what a waveguide is. This handsome piece of precision brass metalwork is a piece of microwave waveguide.

Waveguide

Waveguide is used instead of coaxial cable for signals in the GHz range. Long runs of coax are bad news for GHz waves because at those frequencies coax is both lossy and prone to short circuiting. Waveguide, on the other hand, has very low loss, can't short circuit, and also does a great job of keeping the signal separated from the outside; less noise in the signal, less interference in other gear from the signal.

There's no reason we couldn't make waveguide for, say, HF signals, but you won't find any. Waveguide is (usually) one-half wavelength wide, so a 20-meter rectangular waveguide would be 10 meters wide by 5 meters high. The expense and inconvenience of such an object would be completely unnecessary – coax is perfectly useful at those frequencies.

However, the essence of waveguide is simply that it presents a refracting boundary to a radio wave. It doesn't *have* to be metal – it could have walls that consist of the Earth's surface on one side and the ionosphere on the other, and that's exactly how Earth-ionosphere propagation works. The Earth and the charged fields in the ionosphere create a waveguide for these gigantic waves, something like a giant fiber optic cable, and the waves are guided around the Earth between the ionosphere and the Earth.

Earth-Ionosphere Waveguide Propagation

It's all the juicy goodness of a ground wave only without the lossy ground component – and therefore, lots more distance.

Waveguide propagation also looks a lot like skywave propagation, but it's really different. For one thing, most of the time, it can't possibly be skywave propagation – it's happening *way* below the LUF, the Lowest Usable Frequency. LUF's seldom drop below the range of 100's of kHz, and more often are in the low MHz range. Waveguide propagation occurs mostly in the 3 to 30 kHz range! 3 kHz is a wavelength of about 90 <u>kilo</u>meters or 55 miles – just about the height of the D layer or the bottom of the E layer, depending on conditions.

It also can't be skywave propagation because with waveguide propagation there's no "skip zone" – no area between the transmitter and wherever the skywave signal comes down out of the sky where the signal cannot be heard.

Very reliable long-distance communication is possible with this mode, and, indeed, the military regularly uses VLF – Very Low Frequency – waveguide transmissions for some navigation services (though those are falling by the wayside thanks to GPS) and to communicate globally, particularly with submarines.

Because of its inherently low data rate, there's no practical voice communication on these frequencies, nor any high-speed data. Think in terms of Morse Code at *very* slow speed – 15 second dits, 30 second dahs – or FSK (Frequency Shift Keying) at almost as slow a speed.

Much like the WSJT modes, such as JT65 and FT8, transmission and reception must be tightly coordinated via very accurate clocks, but in the case of VLF, the transmitter and receiver frequencies also must be locked to the same clock – typically a GPS signal.

As you might imagine, the transmitting antenna arrays needed are usually very large indeed. There's a 24.8 kHz Naval Radio Station array not far from my home in Western Washington that occupies the better part of a small valley, with wires strung between towers on the mountains on either side of the valley floor. All total, there's about a mile-and-a-half of wire in the antenna proper, which is near the floor of the valley, and some 13 miles of supporting cables that also act as a big capacitance hat for the antenna.

Big-time VLF also depends on lots and lots of watts. That station near my home runs around 1.5 megawatts, about 1,500,000 times our maximum authorized power on our 2200 meter band at 135 kHz. Obviously, our lowest authorized frequency is far above the 30 MHz top of the waveguide propagation range.

However – hams are unstoppable and, aha, there's a loophole! The ITU (International Telecommunications Union) never allocated any frequencies below 9 kHz to anyone so some hams have been experimenting in the sub-9 kHz band. We don't have any allocated ham bands down there, but since, at least in the world of legal theory, no license is necessary to operate in those bands, the hams have "assumed the authority." They're typically using l-o-n-g wire antennas hooked up to balloons or kites, and have had some success stories even running tiny amounts of power in the microwatt to milliwatt range. Here's what the CEPT countries' Electronic Communications Committee reported in 2015:

"Radio amateurs in several CEPT countries have utilised VLF spectrum for amateur experimentation. In some countries a formal variation to their 'amateur licence' was required; in others no authorisation was required as spectrum below 9 kHz is unregulated. For example, German amateurs chose several spot frequencies e.g. 8.97 kHz, 6.47 kHz and 5.17 kHz for technical convenience for their experimentation. Recently a quantitative field-strength estimate has also been conducted demonstrating that amateur stations are unlikely to cause harmful interference to lightning locator systems in the band 8.3 – 9.0 kHz, given their achievable radiated power levels in the microwatt or low milliwatt range. In the United Kingdom, following a compatibility assessment by the regulator, the band 8.7 - 9.1 kHz has been available to amateur licensees for experimental use on a case by case basis. Countries in other ITU regions have also hosted amateur activities on sub 9 kHz frequencies, notably the United States, Australia and Japan.

Recently an amateur signal on 8.971 kHz with an effective radiated power of circa 150 micro Watts has spanned the Atlantic Ocean, from North Carolina in the United States to the United Kingdom a distance of approximately 6194 km. A steady, GPS-locked carrier at 8.971 kHz was transmitted between 0000 and 0600 UTC and sophisticated digital signal processing (DSP) software was used to detect the transmission under both night-time and daylight propagation conditions at the receiver in the UK."

Those hams experimenting in the low frequency bands are known as "lowfers." Yes, really.

Chapter 3 -- Line-of-sight Propagation

Line-of-sight propagation sounds like it would be a fairly simple proposition. "See receiver, talk to receiver. Don't see receiver, don't talk to receiver. Next topic!" In practice it is seldom that simple. It would not be inaccurate to say that "line-of-sight is never line-of-sight!" That usually works in our favor.

We'd typically expect any of our signals with a wavelength of 2 meters or shorter to be limited to line-of-sight propagation. For those VHF, UHF, and higher frequency signals, the only regular exceptions would be auroral propagation, meteor propagation, and ionospheric propagation made possible by very unusual conditions. Really, even Earth-Moon-Earth propagation is more or less line-of-sight. It just happens to be a line that bounces off the Moon.

By pure "line-of-sight" logic, if a receiver is beyond the visual horizon, we shouldn't be able to communicate with it with, for instance, VHF. However, even in the simple model that we learn for the Technician exam, that isn't *quite* true. The radio horizon is about 15% farther away than the visual horizon, because air diffracts radio waves a bit more than it does light.

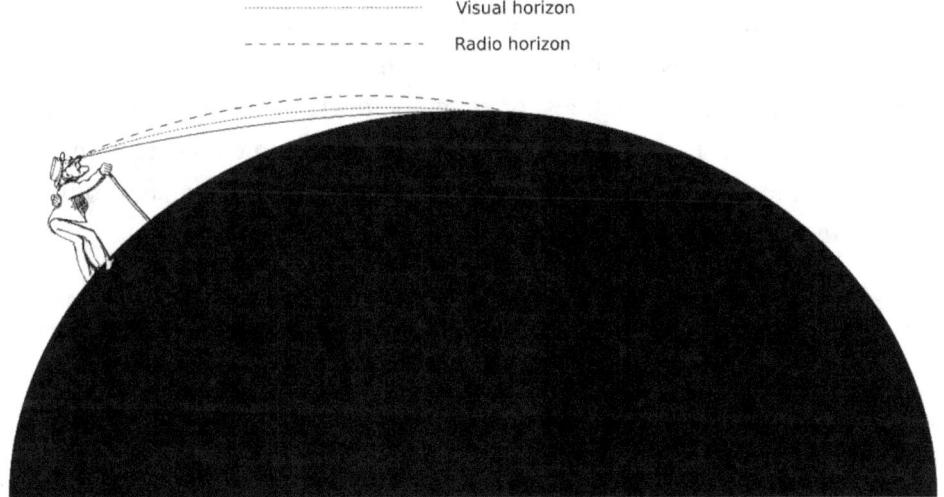

If the visual horizon is 3 miles away – which is about right if you're on perfectly flat ground or looking out to sea – then the radio horizon is about 3.5 miles away. By the way, that visual horizon is a bit farther away than the "real horizon" – the horizon you'd locate by extending a perfectly straight line from your eyes to the point the line ran into the Earth – because of the same effect.

Obviously, if we can raise our antenna higher, we can extend that horizon. If we can gain just 100 feet of altitude, we extend that radio horizon way out to 14.5 miles. Of course, we accomplish exactly the same thing if we raise the an object just over the horizon 100 feet – in other words, if we stick the receiving antenna on a 100 foot tower or, for even more altitude, a tall hill. One of my favorite local repeaters is on a local hill that's about 1500 feet high. That extends that repeater's radio horizon to about 56 miles. No wonder I can hit that machine with my punky little 5 watt handheld from 33 miles away.

With just a little help from circumstances, we can extend that horizon, or even skirt obstacles like buildings and even mountains.

Terrestrial Scatter

If we're on dry land, there's almost always a lot of stuff around that can reflect short wavelength radio waves. Buildings, fences, brick retaining walls, trees, water tanks, semi-trucks, mountains. Most of the time we'd very much prefer to avoid all those reflections, since they tend to create multipath distortion. Sometimes, though, they're actually helping our signal propagate farther than it otherwise would.

Remember, there's no such thing as a unidirectional antenna. Whether the antenna you're using is the rubber ducky on a handheld, a Yagi, or even a parabolic dish, it's transmitting over some range of directions. It might be a 20 degree wide range, 45 degrees, or even 360 degrees. On most places on Earth, that means our signal is going to encounter multiple objects in multiple directions, and that's going to mean our signal has multiple chances to make it to its destination.

Consider it this way. Here's the layout of part of my house.

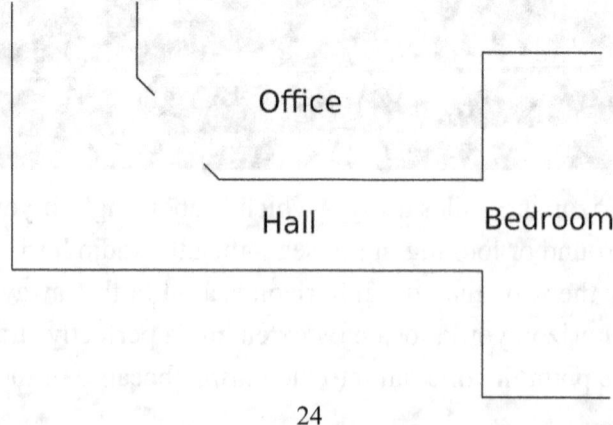

Office

Hall   Bedroom

If it's early morning and I'm in the office, I keep my office door closed. Otherwise, the light shines into the bedroom and causes spousal unhappiness. How can that be? If anything is line-of-sight, it's light, right? The only direction for the light to take out of my office is a full 135 degrees away from the direction it needs to go to get to the bedroom.

Well, obviously, the light is scattering off the walls to reach its destination. Many times our signals will do the same thing. Plenty of hams and more than one TV news station van technician have creatively used terrestrial scatter off a tall building to beam their feed back to the mother ship when they didn't have a direct path to the receiver.

The Fresnel Zone

As mentioned earlier, terrestrial scatter does not usually work in our favor, due to the multipath distortion it can create. There's actually a mathematical way to model the zone that needs to be relatively free of obstacles for our signal to be useful at the receiver end. That model is referred to as the Fresnel (freh-NELL) zone.

For any given transmitter/receiver pair operating at a given frequency, there is a zone between the pair that must be fairly free of obstruction to produce an optimum signal at the receiver end. The zone's shape resembles an ovoid or a zeppelin. It stretches from one end of the path to the other and its radius is dictated by the distance between the transmitter and receiver and by the frequency. Lower frequencies create a larger Fresnel zone.

In the picture above, we have a happy signal path – the Fresnel zone is clear of obstructions.

If something reflective impinges on that Fresnel zone, though, signal quality degrades.

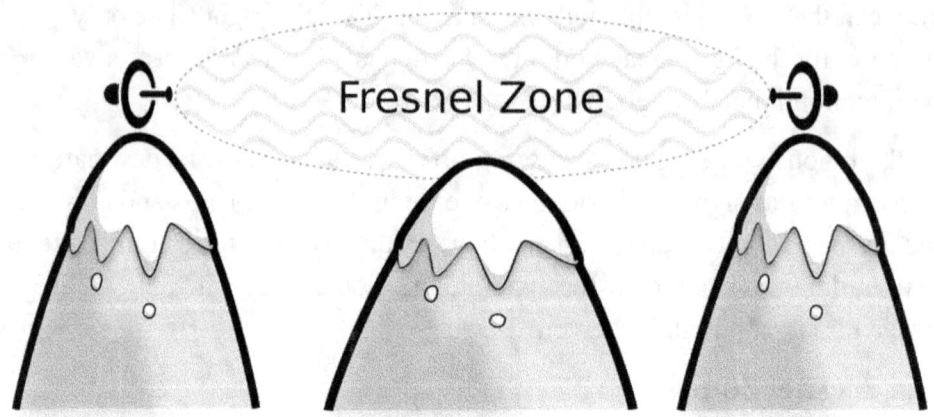

In this picture, that mountain in the middle is going to seriously impede the transmission, even though it's shorter than the other two and there's a perfectly clear line between the two antennas. Put simply, it's going to create an interference pattern that cancels part or all of our signal.

We can calculate the radius of the Fresnel zone at any point along the path with a fairly simple formula:

$$Radius = \frac{1}{2}\sqrt{\lambda d}$$

λ (lambda) is the wavelength of the frequency we want to use and d is the distance from either the transmitter or receiver, whichever is closer. The rule of thumb is that 60% of the Fresnel zone radius needs to be clear at any given point, so for a practical number, the full formula would be:

$$Radius = \frac{1}{2}\sqrt{\lambda d} \times 0.6$$

Here's how this gets used in the real world. First, let's be clear that you're not going to need to calculate the Fresnel zone that extends from your car's mobile radio antenna to your local repeater. That path is either going to work or it isn't. But let's say we want to build a permanent auxiliary transmitter that will send signals to a repeater up on top of a local mountain. We do a path study on one of several computer based tools that can do that sort of thing (there's a free one at https://www.scadacore.com/tools/rf-path/rf-line-of-sight/) and we find something like this:

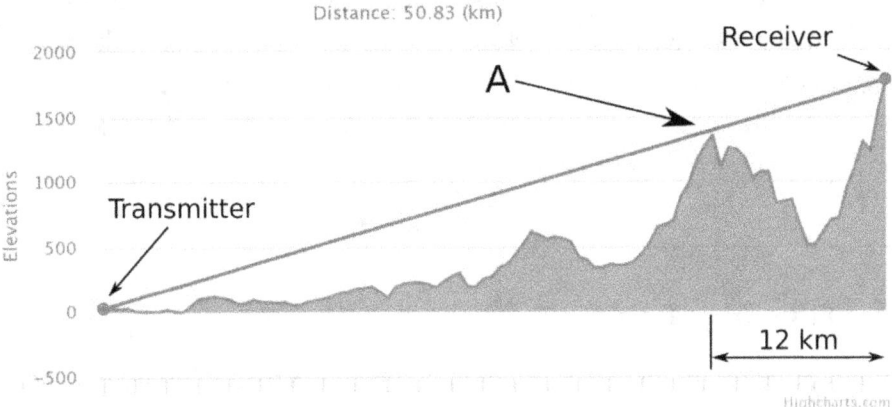

The path study shows us the elevation at each point of the path, as though we were seeing a cross section of the mountains along the signal path. Everything looks great right up until we get 12 km from the receiver at point A, where we figure out we have 20 meters of clearance over the top of that peak. What frequency do we need to use to put a solid signal up to the receiver?

Our club member Bob has been a ham since forever and somewhere in his garage he's pretty sure he has a perfectly fine Heathkit 2-meter radio that he'd be happy to donate to the club for the effort. Good deal? Let's just check that path. We plug in our numbers, using the distance from the nearer of the two radios, in this case, the receiver:

$$Radius = \frac{1}{2}\sqrt{2 \ meters \ \times 12000 \ meters} \times 0.6 = 47 \ meters$$

Uh oh. We only have 20 meters of clearance at point A, so that peak is going to seriously invade our Fresnel zone. This isn't going to work well unless we budget for a taller tower for either the transmitter or receiver. Sorry, Bob!

How about using the 440 MHz band? At that frequency, the radius of our Fresnel zone is still 27 meters. Better, but definitely not optimum.

If we bump up to the 902 MHz band, now the Fresnel zone radius is 19 meters. Whew, we just skate over the top. For safety's sake, though, we'd probably go on up to the 23 cm, 1240 MHz band, where our Fresnel zone radius is only about 16 meters at point A.

I hope it won't add too much complication to this topic to mention that I've presented you a simplified model of Fresnel zone theory. In the full Fresnel zone model, there are three significant Fresnel zones, each a little larger than the other. I've described the inner Fresnel zone, known as Fresnel zone 1 or F1. F2, the next larger zone, is a zone in which an obstruction in the path does not produce a phase difference between the direct and reflected signal. In zone F3, the direct and reflected signals are once again out of phase. The formula for any Fresnel zone radius n is:

$$Radius_n = \frac{1}{2}\sqrt{n\lambda d} \times 0.6$$

Knife Edge Diffraction

Another propagation phenomenon that can take your signal to unexpected places is knife edge diffraction.

Knife edge diffraction occurs when a signal encounters a sharp edge of some sort – for instance, the edge of a building or the top of a (steep) mountain ridge. The signal seems to bend around the building or over the top of the ridge and back down the other side.

Way back in the 1600's, Christiaan Huygens, physicist, astronomer, mathematician, and inventor of the pendulum clock, came up with a theory that explains just how knife edge diffraction occurs.

In Huygens' model, light – he didn't know about the other electromagnetic waves, just light – was an "advancing wavefront." Each point on that wavefront is the source of a spherical "wavelet", and the fronts of those spheres become the new wavefront.

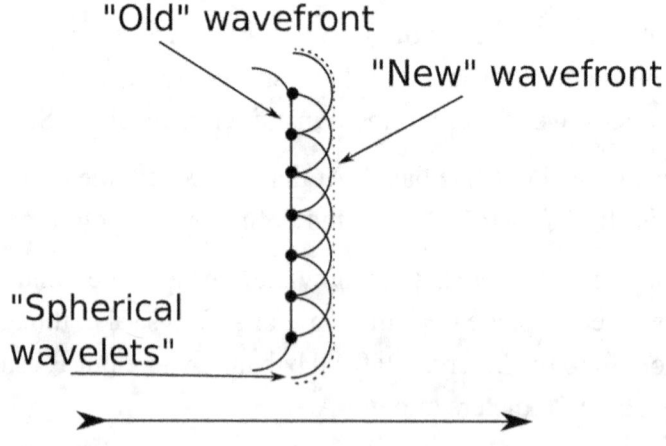

You can see that as the wave advances, it's going to spread wider and wider. (Of course, in reality, it's a three dimensional wave, so it's spreading wider, higher and lower.)

If our Huygens wavelets encounter a sharp edged obstacle, of course some of them get blocked. The rest, though, continue to spread out and, if the geometry's on our side, our signal seems to roll right down the other side of the mountain. Ah, but there's no such thing as a free lunch in this ham radio game!

You might remember I mentioned in the previous section that sticking an object into the Fresnel zone of a signal scrambles the phase relationships of the received signal. As you might guess, knife edge diffraction is related to the Fresnel zone, and sure enough, what gets created on the other side of the hill is an interference pattern. The signal is present in some places, and has cancelled itself out in others.

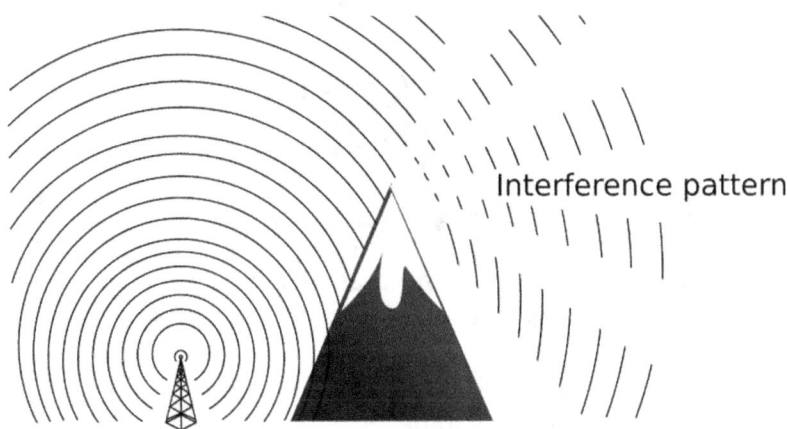

Interference pattern

We can predict whether our signal will make it over the hill with this formula, which is directly related to the formula for the Fresnel zone. The last formula told us the size of the radius of the Fresnel zone that needs to be unimpeded to avoid signal loss. This formula tells us (indirectly) how bad the signal loss is going to be if that zone does get impeded:

$$v = h\sqrt{\frac{2(d_1 + d_2)}{\lambda d_1 d_2}}$$

In this formula, d_1 is the distance between the transmitter and the obstruction along the line of sight, d_2 is the distance between the receiver and the obstruction along the line of sight, h is the height of the obstruction above the line of sight and λ is the wavelength. (We can enter 0 for the height if the ground is flat, and use ½ the distance for d_1 and d_2.)

The result of the formula, v, is the "Fresnel Diffraction Parameter." That will tell us whether the signal will make it over the hill. If it comes out to -1 or less, we're 100% in business with 0 loss – or close to it, anyway. For it to be -1 or less, though, h, the height, has to be a negative value. In other words, it's a hole in the ground or a canyon – or the height of our antenna above the terrain.

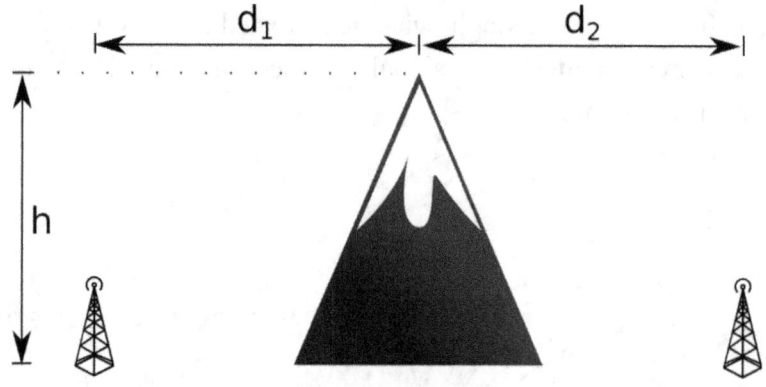

Most of us realize early on in our ham career that antenna height is our friend, and this really shows why "height is might."

The next chart shows the approximate amount of loss we can expect at each value of the Fresnel Diffraction Parameter:

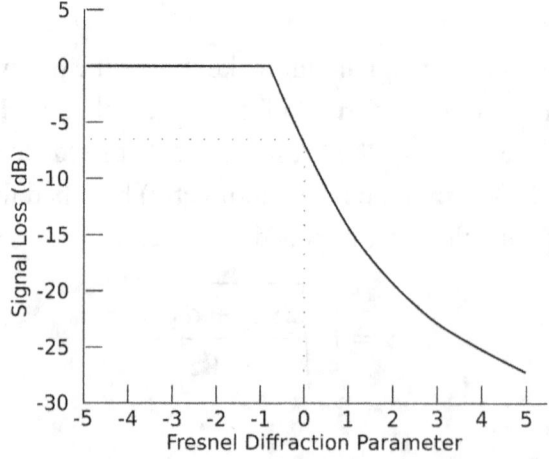

Notice that at a Fresnel Diffraction Parameter of 0, we still get 6 dB of loss. In English, that means there's going to be 6 dB of loss across perfectly flat ground between an antenna at zero height and another antenna at zero height. 6 dB means we're delivering ¼ of the signal we could deliver with a little height. Why is that? Fresnel zone theory tells us any obstruction within the Fresnel zone will degrade the signal. At zero height, the entire lower half of the Fresnel zone is being impeded by the Earth!

If we stick each of our antennas on 20-meter tall towers, here's how it works out. We'll assume we're sending a 70 centimeter signal 2 kilometers over, conveniently enough, flat ground, so that our HAAT (Height Above Average Terrain) is the same as the tower height.

Now h becomes -20, and d_1 and d_2 are each half of that 2 km hop.

$$v = h\sqrt{\frac{2(d_1 + d_2)}{\lambda d_1 d_2}} = -20\sqrt{\frac{2(1000 + 1000)}{0.7 \times 1000 \times 1000}} = -1.51$$

In this formula, as wavelength increases, the value of that fraction under the square root sign gets smaller. That means for negative values of h, the value of v gets less negative as wavelengths get longer. For instance, in the above formula, if we plug in 2 meters instead of 70 centimeters, v = -0.89. For positive values of h – which means something's in our way -- the value of v gets lower. Translated: if you're trying to get over a hill, use a longer wavelength. If you have a clear shot, use a shorter wavelength.

Tropospheric Scatter

The layer of the atmosphere we're breathing is called the troposphere. It extends up to about 30,000 feet above sea level.

The troposphere is a turbulent place! It's full of pockets and layers that are a different temperature or humidity than the surrounding air, layers that are blowing this way and that, and whirls and eddies. If you've ever had a turbulent airplane ride, you've experienced it. All that adds up to pockets of different densities, and whenever we have changes in density we have the possibility of diffraction.

The troposphere doesn't have much effect on HF signals – those wavelengths are just too big for the little disturbances in the troposphere to have much effect. For VHF and UHF, it's a different story. Those frequencies get refracted by all those "air pockets." In fact, they get refracted rather predictably – tropospheric scatter ("troposcatter" to its fans) is a quite reliable propagation mode.

Here's how it works. This is our usual "high school physics" picture of us trying to send a signal over the horizon.

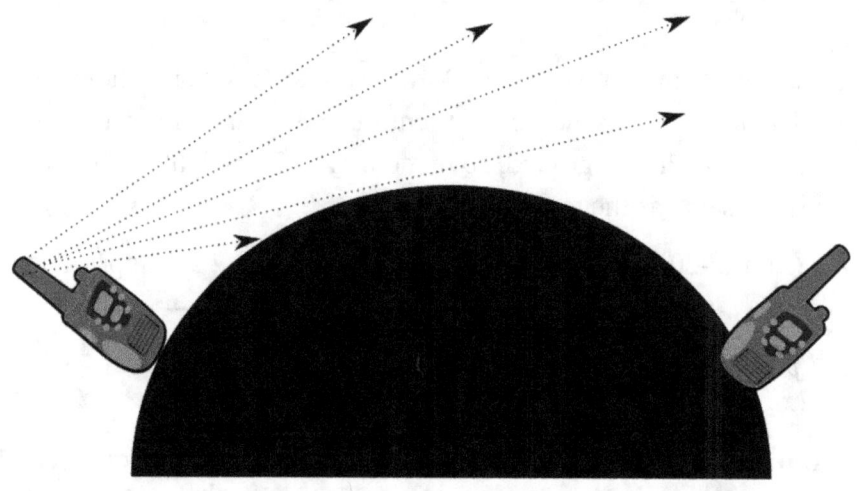

Our signal is just shooting off into space and totally missing that over-the-horizon receiver.

However, with a little tropospheric scatter, the picture might look at lot more like this:

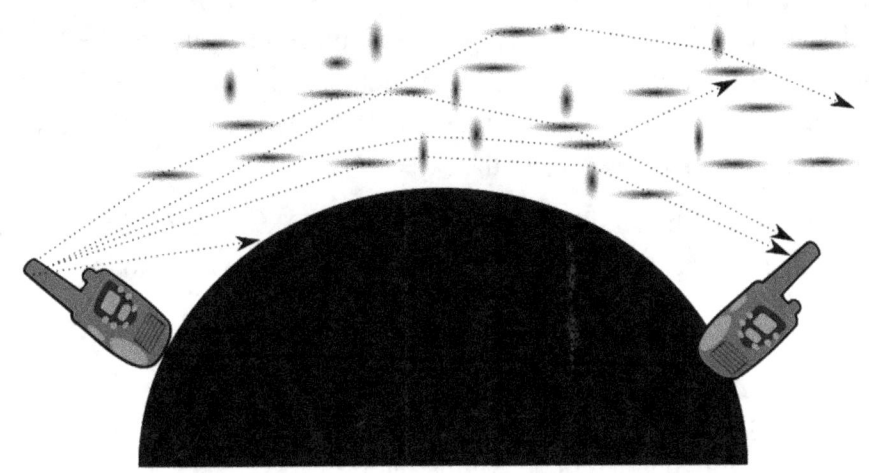

Now our signal is encountering a bunch of "air pockets" and getting bent just a little bit by each one. Some of our signal makes it "over the hill," and some of it gets shot off to space even sooner, but if we can put enough signal into the receiver, we've achieved contact.

Here's a real-life example of what is most likely tropospheric scatter at work. This is the path study from our house to our club's repeater.

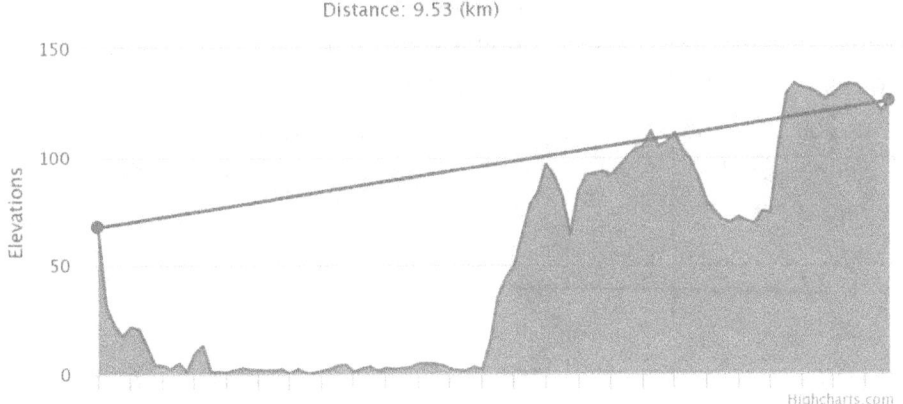

That's us on the right. You can see, the line-of-sight path barely leaves our antenna before it smacks into a couple of hills that are about 10 meters higher than us, and then there are two more peaks that poke into our Fresnel zone. On paper, it doesn't look like anyone's going to hear from us on the Sunday night 2-meter net! Yet, we're very reliably on there every week that we're in town. Knife-edge diffraction off those hills? Unlikely because of the proximity of those hills on the far right, which puts our Fresnel Diffraction Parameter at a whopping 22.5. It's much more likely that tropospheric scatter is bouncing our signal down from the sky.

Troposcatter is much loved by fans of VHF/UHF DX. It's far more reliable than sporadic E. Bob Atkins, KA1GT, runs a web site devoted to VHF/UHF/Microwave information for hams

(http://www.bobatkins.com/radio/index.html)

where you can also find all the math to calculate the possibilities in a particular situation, but here's one of his samples:

As an example, take two stations with a capability of 100 watts output on 432 MHz, each using a single 18 dBi gain Yagi with a receiver noise figure of 1 dB and having 1 dB line loss on both transmit and receive. If both stations have an unobstructed view of the horizon (take-off angle of zero degrees), they should be able to communicate on CW over a range of 650 km. This is considerably greater than most station operators would guess, and suggests the unrealized potential of many stations.

650 km – about 400 miles – is a *long* way over the horizon.

You might have guessed from the previous example that success at using tropospheric scatter for DX is all about putting a lot of signal in a very specific direction. A highly directional antenna, aimed right at the horizon between you and that other station is key. Because the actual bending is much, much less than shown in the illustration above, the higher the antennas, the better.

Chapter 4 -- Tropospheric Ducting

In normal conditions, as we go higher in the troposphere, the air temperature gets cooler. A **temperature inversion** occurs when there is a layer of warm air sandwiched between moist air at the surface, and cool air above. The differences in density at the boundaries of those layers can refract radio waves.

There are really three varieties of tropospheric propagation that we could lump into "tropospheric ducting."

Tropospheric Bending

The first occurs when we have a layer of cool, moist air at surface level, topped by a layer of warm, dry air with only a weak or moderate inversion – only a small difference in temperature. In that case, we'll get what's called "super-refraction" or just "tropospheric bending." Signals leaving the transmitter at a high angle of incidence relative to the ground go right through the inversion and off to space. Signals with a lower angle of incidence are refracted off the inversion, then bounce off the ground – or not. Between the radio horizon and where the signal comes back down there's a "skip zone." Because there's a lot of loss both at the weak inversion and when the signal bounces off the Earth, this is typically a shorter range form of ducting propagation.

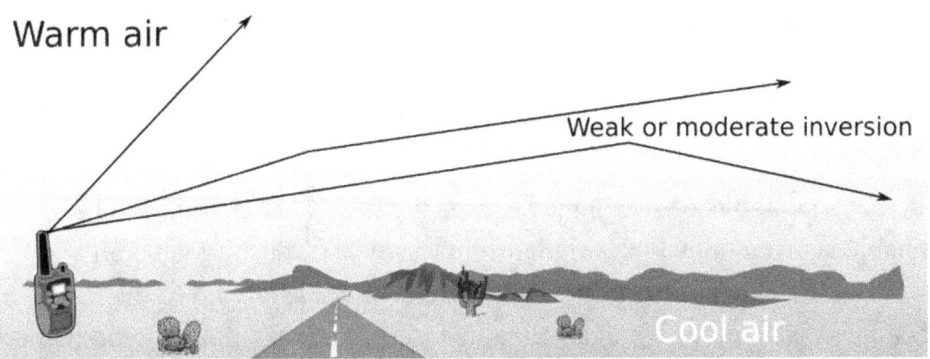

Tropospheric super-refraction

"Normal" Ducting

With a greater difference in temperature between the layers, the refractive index of the boundary becomes stronger and what we usually think of as "normal" ducting can occur. Now the inversion layer is providing a much stronger reflection back to Earth, so this form of ducting can provide ranges of longer distance, particularly if the signal is traveling over the sea where the earthbound side of the duct provides a much better reflector. Note that in this type of tropospheric ducting, the signal will typically be able to be received at most (if not all) points along the signal path.

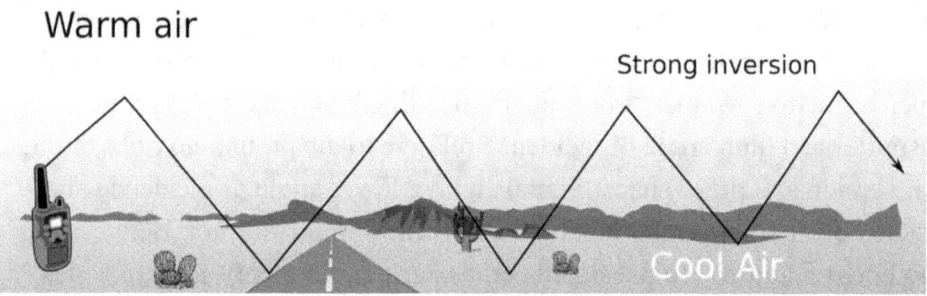

"Normal" ducting

Elevated Ducting

Yet another type of ducting is known as "elevated ducting." This can occur when the top of the inversion is relatively high and temperatures close to the surface are relatively mild. In these conditions, moist air in the inversion can accumulate near the top of the inversion, creating an elevated duct that "traps" the signal.

While occasional bits of signal may escape the duct's lower edge and be detectable at ground level, the signal from this type of ducting generally won't be receivable until it comes out of the duct at the far end of the inversion, or is received by an antenna in a location like a mountaintop that is poking up into the duct.

Depending on the reflectivity of the upper and lower layers, and, of course, on how much area the inversion covers, elevated ducting can transport signals for very long distances in the 1,000 mile and up range.

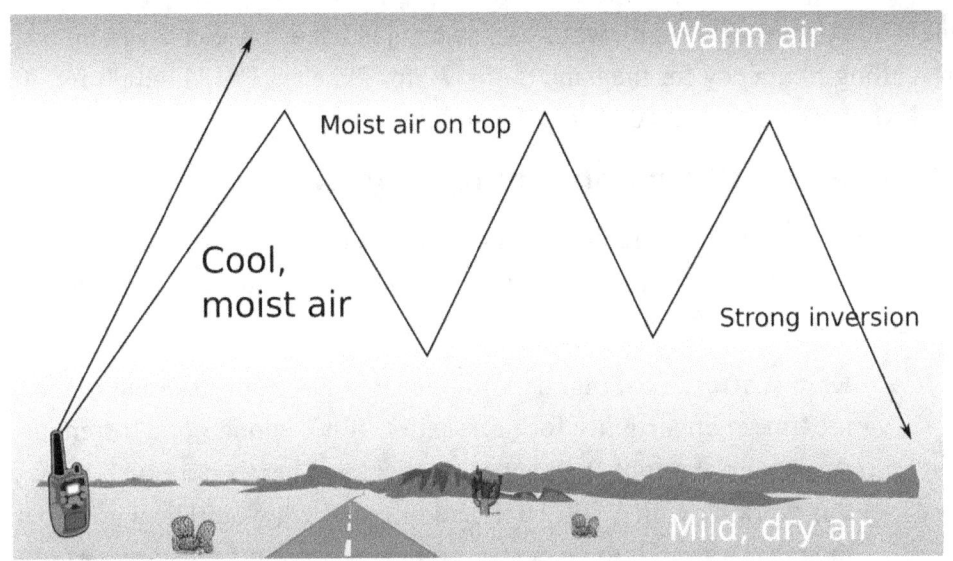

Elevated Ducting

Generally speaking, tropospheric ducting will be most common in the summer and fall on still days when you're in a high pressure system. Some areas of the country will almost never experience tropospheric ducting, because their terrain is so mountainous that the updrafts and downdrafts created by the mountains tend to break up the warm air layer.

The most common place to find lots of tropospheric ducting is over the ocean. Its early November now, and I just took a look at the Hepburn map for today. (William R. Hepburn is a Canadian VHF DX listener who publishes daily hour-by-hour tropospheric ducting forecasts at:

http://www.dxinfocentre.com/

His "Hepburn maps" are even a topic on the current Extra exam!)

Today, there's ducting that covers the Pacific from the West Coast clear out to Hawaii, though it's pretty sketchy after you get halfway to Hawaii. Still, in better conditions, contacts have been made, via ducting on UHF, from Los Angeles to Honolulu.

In some parts of the world, such as the Mediterranean, ducting conditions can remain stable for months at a time, allowing residents to regularly watch television signals from hundreds of miles away.

Your best VHF/UHF setup to take advantage of tropospheric ducting will most likely be a horizontally polarized beam antenna with a nice low takeoff angle hooked up to as many watts as you can lay hands on. Set the radio on the calling frequency for the band (146.520 for 2 meters FM, 446.000 for 70 cm FM, or 144.200/432.100 for SSB) and go for it!

Other Forms of Tropospheric Propagation

Hepburn's "DX Info Centre" site, which touches on all forms of DX, lists a few more forms of tropospheric propagation that can open up long distance contacts that wouldn't otherwise be possible.

- **Rain scatter:** That big, dark rain front you see approaching might present an opportunity for rain scatter. If it's enough of a torrential downpour, it might even act as a reflector. There's not much hope of this working with VHF. UHF and microwaves would be a good bet, though. (After all, microwaves bouncing off rain fronts are part of how weather radar works.)

- **Hail scatter:** Similar to rain scatter. (Stay safe; hail is often accompanied by lightning.)

- **Sleet scatter:** Sleet is not a common occurrence in most of the US, but if you're in the Northeast you know you see a few days of it most years. Keep an eye on the weather channel for opportunities for rain, hail or sleet scatter.

- **Lightning scatter:** This one is iffy. It's "seldom reported", according to Mr. Hepburn, but at least in theory it's possible. Each lightning strike leaves behind a column of highly ionized air, much the same as meteors leave in the E layer as they superheat. At best this is going to give you a brief burst of reception, since the storm's turbulence is going to dissipate the column of ions quickly.

- **Aircraft scatter:** Big shiny pieces of aluminum in the sky! Of course, they're going to reflect VHF/UHF/microwave signals. Before you aim your parabolic dish and pump 1500 watts of UHF at the airplane, do consider that it's not only seriously illegal to interfere with navigation systems, it's the sort of experiment that could create a disaster. However, if you happen to get a brief bit of long distance propagation that you can't otherwise explain, it just might be a passing jet helping you out.

Chapter 5 -- Ionosphere Basics

If all your experience with ham radio has been in VHF and UHF, and you're just now making the leap into HF, you're about to enter a very different world. The weather here on Earth can have dramatic effects on VHF/UHF (and higher) signals, while, with rare exceptions, the weather on the Sun has almost nothing to do with our ability to communicate by those frequencies. With HF, it's exactly the opposite. Terrestrial weather has little effect on HF, while solar weather has enormous effects.

Aside from a passing reference to "sporadic E", the Technician exam basically waves in the general direction of the ionosphere and notes that it's up there and sometimes it bounces radio waves. Let's face it, the vast majority of Technician licensees spend their radio time in the VHF/UHF bands and don't deal with the ionosphere.

Once you get your General License and start using the HF bands, your relationship with the ionosphere changes. Indeed, more than one ham has gotten quite emotional – one way or the other -- about the condition of the ionosphere! The more you understand about the ionospheric events that affect HF propagation, the more you're going to enjoy those bands.

Fair warning: Not all questions will be answered, because at least for now, there are far more mysteries than certainties in the science of the far reaches of the atmosphere. Consider this: four large, complex systems interact to create or destroy radio propagation via the ionosphere. They are the Sun, the Earth's magnetic field, the lower atmosphere, and the ionosphere itself. Even if we completely understood *everything* about those four systems, we'd *still* have a tough time making accurate predictions, simply because of the complexity involved. That's something to consider before you place 100% faith in propagation predictions, whether they be for great or dismal conditions.

The ionosphere is the second highest part of Earth's atmosphere. The highest is the exosphere, which is really just the transition to deep space.

The ionosphere starts about 30 miles above the ground and continues way on out to about 1000 miles above the ground. The air gets very, very thin up there; the molecules get farther and farther away from each other. The lowest layer of the ionosphere is about 1,000,000 times less dense than the air you're breathing.

The Sun is constantly bombarding the Earth with electromagnetic radiation, from infrared (heat) all the way out to x-rays and gamma rays. The part of the radiation that is *ionizing radiation* is energetic enough to *ionize* some of the air molecules it encounters. A molecule gets *ionized* when it picks up a charge by either gaining or losing an electron. In nature, molecules mostly lose electrons.

Most of the ionizing radiation -- ultraviolet and higher frequencies like x-rays and gamma rays -- gets filtered out by the atmosphere before it ever gets down here where we are by the very mechanism just described. Because those frequencies get filtered out by that process, the closer we get to Earth, the less ionization is occurring.

So imagine a molecule of air way up there in the ionosphere. It gets blasted by a stray bit of that ionizing radiation, an electron gets knocked out of its orbit around the nucleus and suddenly finds itself flying around in space with no home. The free electron is packing a negative charge, and the atom it just left is packing a positive charge – it has been ionized.

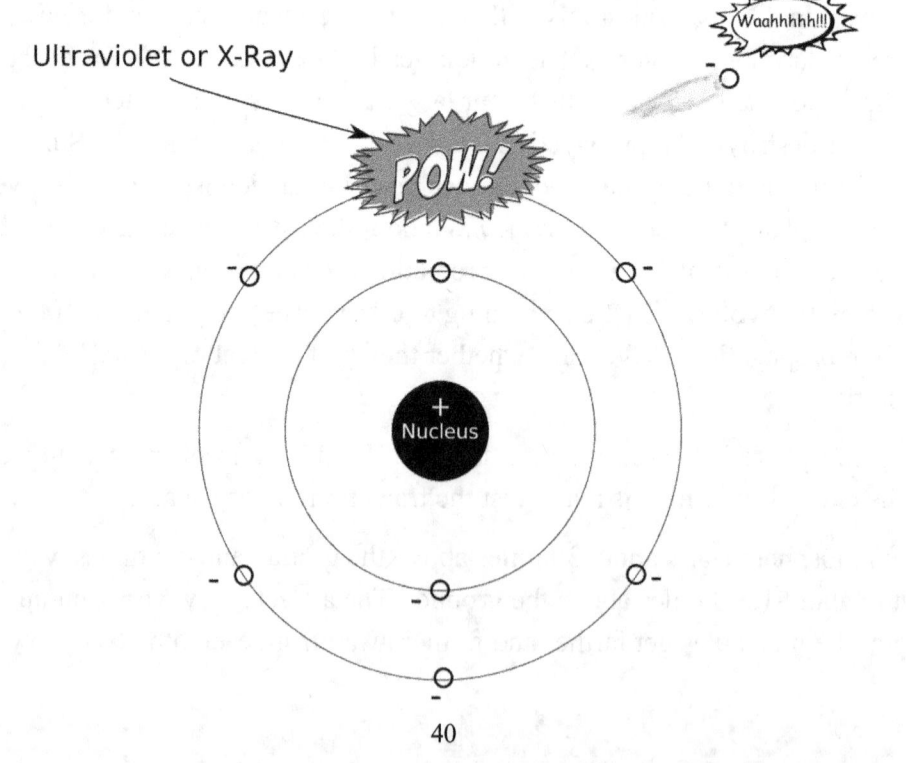

That stray electron would really "like" to hook up with another nucleus, but there isn't one nearby. The electron gets knocked loose *way* out into empty space -- at least, *way* out on the atomic scale of things. It's free floating with nowhere to go. And even if that lonesome electron does find another home, that atom gets hit with another packet of ionizing radiation -- because there's lots of it up there – and, ZAP, another electron is blasted loose. Take uncountable gazillions of those events, average them out, and you have a chunk of upper atmosphere that is no longer electrically neutral. It has a certain concentration of free electrons, and when there are lots of free electrons things get interesting for our radio waves.

The amount of ionization occurring is not constant. The level of this electrical activity in the ionosphere is directly related to the time of day here on Earth, to the season here on Earth, and to events on the Sun; most importantly the level of sun spot activity and the Solar Flux, about which more later.

The rate of recombination, though, is relatively constant at a given altitude, because it's mostly determined by the density of the gas. We have two competing processes going on, then; ionization and recombination. During the day, ionization wins the race. At night, recombination wins. The lower the layer, the faster the recombination.

The amount of time the average electron spends as a free electron before recombining correlates directly with the Lowest Useable Frequency, the LUF. You see, that electron must remain free for at least the duration of one cycle of your signal in order to reradiate that signal. If it recombines with an atom before the cycle is finished, the energy of your signal just gets turned into heat rather than electromagnetic radiation.

"Critical frequency" relates to the density of the free electrons present. Critical frequency is the highest frequency signal, going straight up, that will be reflected by the layer. Higher frequencies than the critical frequency shoot straight through the ionosphere and continue off into space. Higher electron densities reflect higher frequency signals, and free electron density is, in effect, the whole critical frequency game! In fact, the formula for the critical frequency only needs to know the electron density, "*N*" to work:

$$f_{critical}(MHz) = 9.10^{-6} \times \sqrt{N}$$

The critical frequency is not, generally, equal to the MUF, the Maximum Useable Frequency, but they are directly related. The MUF is the highest frequency that will be refracted back to Earth, given the current conditions and the takeoff angle of the signal – about which more in a moment. The MUF will normally be higher because you're not sending your signal straight up, so your signal will encounter more ionosphere to refract it.

The Anatomy of The Ionosphere

While we call it the iono*sphere*, we know now that the charged layer forms more of an off-center egg shape around the Earth, with the thinnest layer over the nighttime side of Earth.

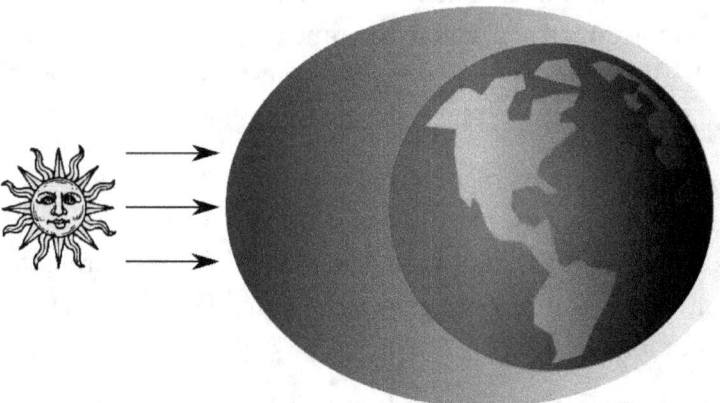

(The scale of most of this book's illustrations is wildly exaggerated for clarity.)

Of course, as altitude increases, the air gets thinner and thinner, so there are fewer and fewer molecules to ionize – but there is also less shielding from the solar energy that is doing the ionization because there's less atmosphere above to do the shielding. Also, because the air is thinner at higher altitudes, there are fewer ionized atoms with which the free electrons can recombine; in other words, the higher the altitude, the longer the ionization lasts. When the free electron recombines with an ion, we're back to neutral; and neutral atoms don't affect our radio waves. Lower in the atmosphere there are more atoms to ionize, but more atoms with which to recombine. It's a complicated equation! The way it all works out, there are actually more free electrons per cubic centimeter in the uppermost layers of the ionosphere than in the D and E layers.

Our atmosphere consists almost entirely of nitrogen and oxygen. The proportions of those gases change as we go up, with the upper layers containing a higher proportion of oxygen ions. At some levels, atomic nitrogen and ionized oxygen combine to form nitric oxide. So here's what we have swirling around in various areas of the ionosphere:

- Ordinary, "elemental" oxygen, O_2
- "Atomic" oxygen, O
- Ionized oxygen, O^+
- Elemental nitrogen, N_2
- Atomic nitrogen, N (but not much of it – it turns into NO^+.)
- Ionized nitrogen, N^+
- Ionized nitric oxide, NO^+.
- There are even metallic ions floating around up there, left over from meteors that burned up in the ionosphere!

It's a complicated chemical stew, to say the least.

Due mostly to the interactions of altitude, air density, absorption of solar radiation by the higher layers, and the chemical composition at various altitudes, the ionosphere acts as if it is composed of layers. These areas tend to be the ones we like for propagation. We refer to them as layers, but understand; they're much less like layers of a birthday cake than they are like layers of smoke, subject to being disturbed or even completely blown away by various forces.

When Edward Appleton first started researching the ionosphere, he discovered one of these layers and named it the "E layer" because it returned Electromagnetic waves. Later, other layers were discovered and were named the D, F, and F1 and F2. From the ground up, they stack up D, E, F1 and F2. At night, F1 and F2 blend into one F layer and the D layer disappears.

Here's a "traditional" model of the ionosphere, as taught in countless textbooks, and even my own *Fast Track* ham license books:

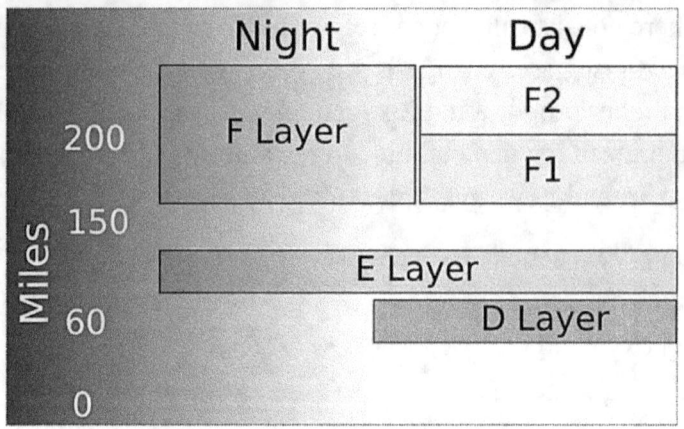

Those nice neat boxes are a useful basic model for passing a license exam, but they really don't reflect reality, and they don't tell us much of anything about how we can work more DX. Instead, the ionosphere contains areas where the electron density increases sharply, and others where it remains more or less the same for some distance. Those variations are what create the "layers."

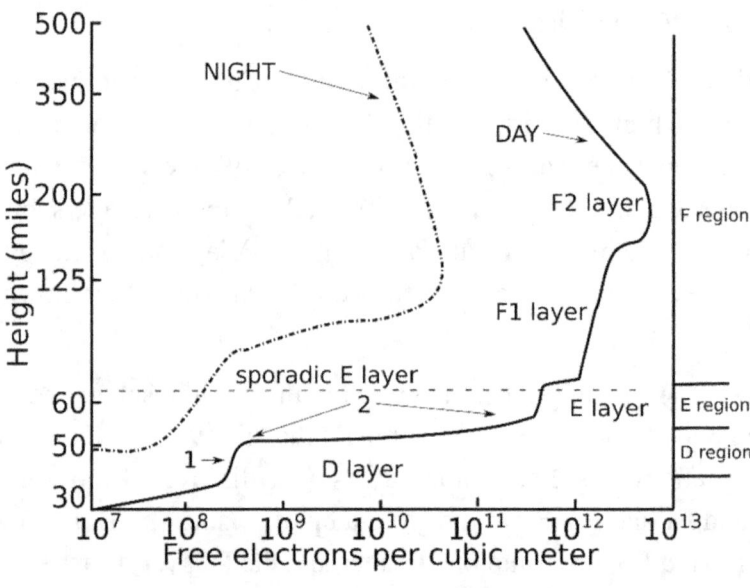

Electron Density vs. Altitude (typical)

When we measure electron density and plot it against altitude, we end up with curves that usually resemble those. To read that plot, notice that free electron density increases along the horizontal X axis, and altitude along the vertical Y axis.

During the day – shown by the solid-line plot on the right of the chart – electron density climbs gradually up to the 40 mile high mark which is the usual location of the D layer. Then there's a "step" in the plot; what physicists call a "steep vertical gradient," marked by the "1" in the illustration above. The electron density doesn't change much for about ten miles at point 1, then increases rapidly over very little space on a "steep horizontal gradient," marked by "2" above. You can see another shorter steep horizontal gradient at the E layer, and another two that define the F layers.

At night, the step that makes the D layer disappears, and the F1 and F2 steps have blended into a single big gradient, known at night simply as the F layer.

You can see that the greatest free electron density is found in the F region, even though the actual air molecules are rather few and far between compared to the lower levels. Even better, for our purposes, is the fact that because those air molecules are relatively rare, there's less of a chance that our helpful free-floating electron will find a positively charged atom and recombine with it. While a free-floating electron in the (again, relatively) densely packed D layer might only last microseconds before recombining, a free electron in the upper reaches of the F layer can last for days. Indeed, it is the long-lasting nature of free electrons in the F layer that makes nighttime skywave propagation possible.

Those steep gradients, the layers where electron density rapidly increases, are the key to the refraction of our signals. Without those gradients, all our signals would blast into space and never come back. Refraction depends on sudden changes in density.

If we're scuba diving and look at a pencil in our hand, everything looks normal even though we're seeing it through water. Of course, the same applies if we're on land. Create a sudden change in density from air to water, though, and we see refraction:

Refraction is what's happening to our radio waves, by the way, not reflection, even though it's convenient to think of it as reflection. Our signals aren't exactly ricocheting off the ionosphere, they're being curved back to Earth, something like this.

Refraction

On the sub-atomic level, our signal is being absorbed and reradiated by free electrons. Each time it gets reradiated, the signal bends a bit back toward Earth. That's simply because of the *refractive index* of the ionosphere, which will always tend to bend signals back toward Earth. For a given frequency, the refractive index decreases while passing from a medium of lower free electron density to higher free electron density. Remember, the free electron density increases with altitude. That's why our signal gets refracted back down to Earth. If the free electron density decreased with altitude, we'd be seriously out of business when it comes to ionospheric propagation.

One major influence on the structure and behavior of the ionosphere is the Earth's magnetosphere. The magnetosphere's shape is the result of the interaction of Earth's magnetic field with the solar wind and, to some extent, the Sun's magnetic field. The classic model of Earth's magnetic field is something like this, with Earth depicted as, essentially, a giant bar magnet.

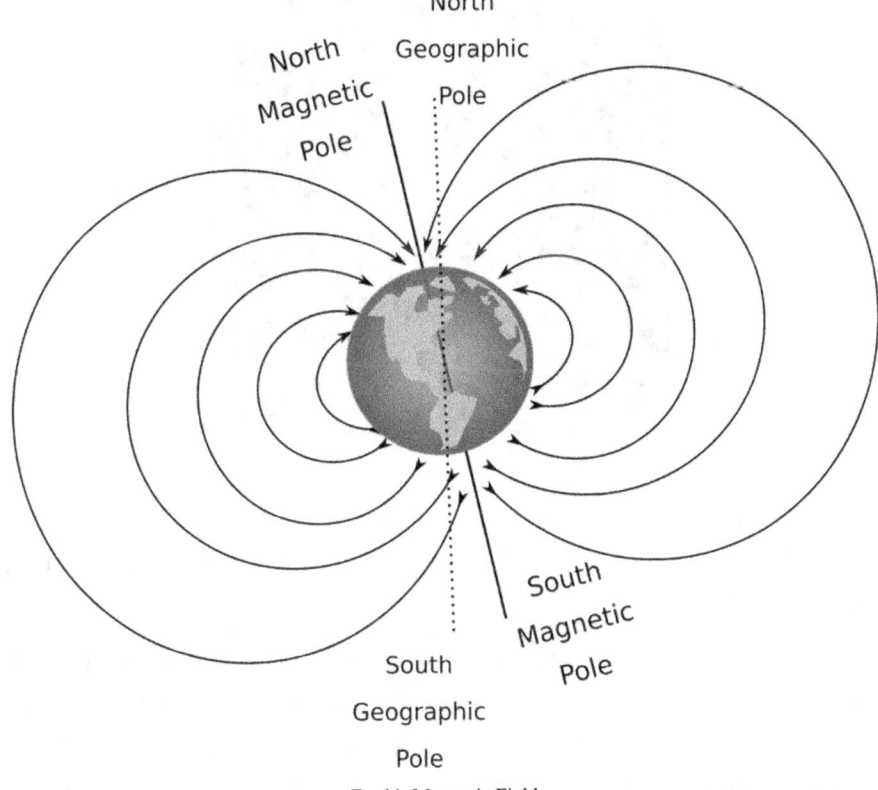

Earth's Magnetic Field

On paper it looks like it's stable and symmetrical as can be. It very much is not.

Even if we left out the effects of the Sun, the field would still be constantly twisting and torquing upon itself because it is the result (probably – we don't know for sure) of what is basically a giant, sloppy electrical generator at the center of the Earth. As the Earth's solid iron core sloshes and rotates relative to the surrounding molten magma, electrical currents are created. Geologists call this the geodynamo. As we know, where there is current, there is magnetism, and in this case it is magnetism that tends to move around a lot; it practically ties itself in knots.

The *real* picture of Earth's magnetic field is a bit messier than the one with the nice symmetrical lines of magnetic force that look like some cosmic giant sprinkled iron filings on a planet-sized bar magnet. This one was created at the Pittsburgh Supercomputing Center.

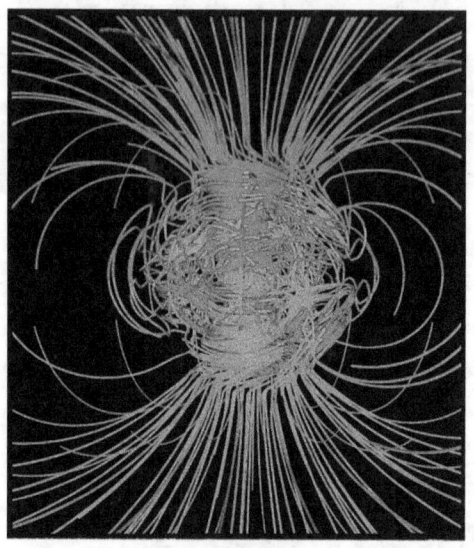

Earth's Magnetic Field

That's the magnetic field. The most convoluted portions of what you see there are below the Earth's surface, but even the surface has some twisted areas, and areas of asymmetry.

The solar wind hits the magnetic field hard – it's traveling close to 1,000,000 mph, and sometimes faster -- and creates the magnetosphere. The sunward facing side of the magnetosphere is called the bow shock area; it's where the solar wind starts to encounter our magnetic field and get diverted, usually, around Earth. On the dark side, it tears part of the magnetic field away in a long, dwindling streak of magnetism and particles.

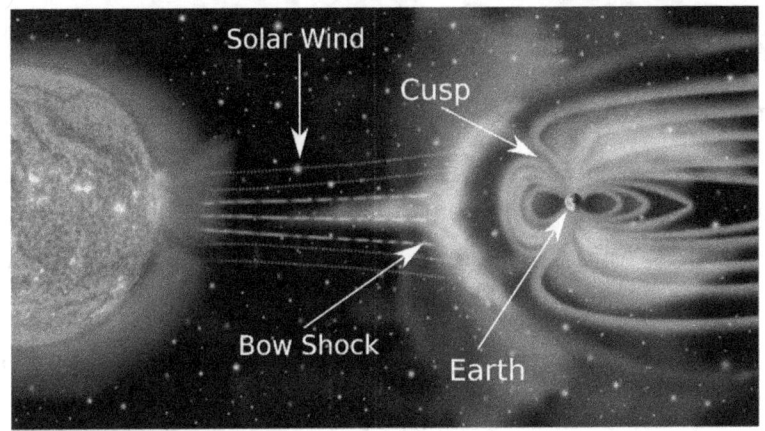

What you see in that picture is not the ionosphere, it's the environment in which the ionosphere exists. In the scale of that picture, the ionosphere is indistinguishable from the surface of the Earth.

Even on "quiet sun" days, the magnetosphere contributes greatly to the asymmetrical "geography" of the ionosphere – at the most basic level, it contributes to that off-center egg shape around the Earth, though that's mostly a function of the Earth casting a shadow on the side of the ionosphere that's away from the Sun.

The geography of the ionosphere as it relates to critical frequency is also, usually, a bit askew from what we'd expect. For instance, let's say it's 0600 in England (UTC) on the day of either the fall or spring equinox.

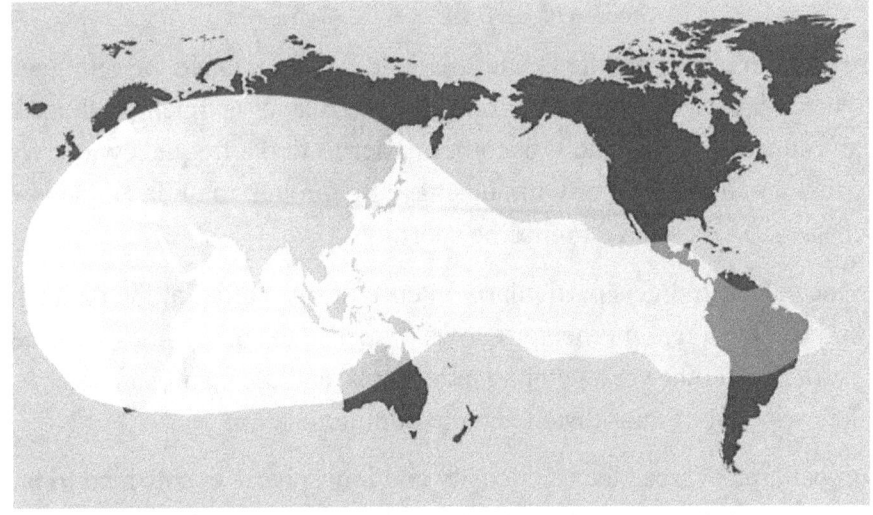

Areas of Highest F2 Free Electron Density

49

If we mapped, in white, the highest critical frequencies of the F/F2 layer all around the globe at that moment, it would probably have a shape that at least vaguely resembled the bright area of that map.

Even though on this day the Sun is firing directly at the Equator, the critical frequency map slumps a little to the South in the Western Hemisphere. That's because it tends to follow the geomagnetic lines of latitude, not the geographic lines.

By the way, the densest areas of the D and E layers at the same time on the same day would look something like this map.

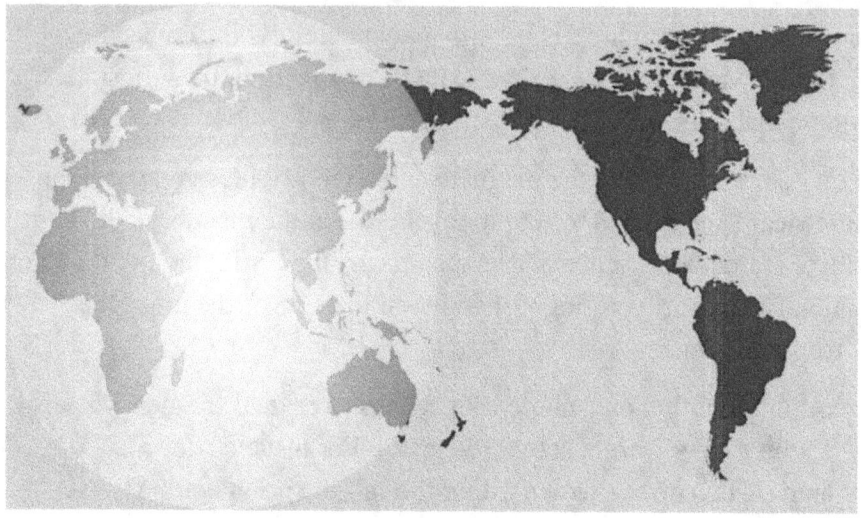

Areas of Highest D & E Layer Electron Density

Where the Sun's shining, the D and E layer are strong. Once the Sun goes down, the D layer is, essentially, gone, and the E layer is much weakened. Its critical frequency drops, and it becomes easier to find a frequency that will pierce the remaining E layer without smashing through the F layer. This is why chasing DX is often a nighttime sport.

When the solar wind gets particularly intense, it rips harder at the magnetic field and can even tear momentary holes in it. The magnetic lines of force twist across each other in a geomagnetic storm, and the result can be a complete wipeout of long-distance radio communication.

Certain patterns of free electron density and ionospheric events tend to be located in particular bands of geomagnetic latitude; the high or "polar" latitudes, middle latitudes, and equatorial or "low" latitudes. The precise definitions of those latitudes, in terms of degrees, are vague. Generally, the low latitudes are said to extend from the equator to 50° north and south. The lower 48 US states are in the upper reaches of the low latitudes. The mid latitudes cover the ten degrees from 50° to 60° north and south; from the south edge of Canada to, roughly, Anchorage, AK. The high latitudes run from 60° north and south to the poles.

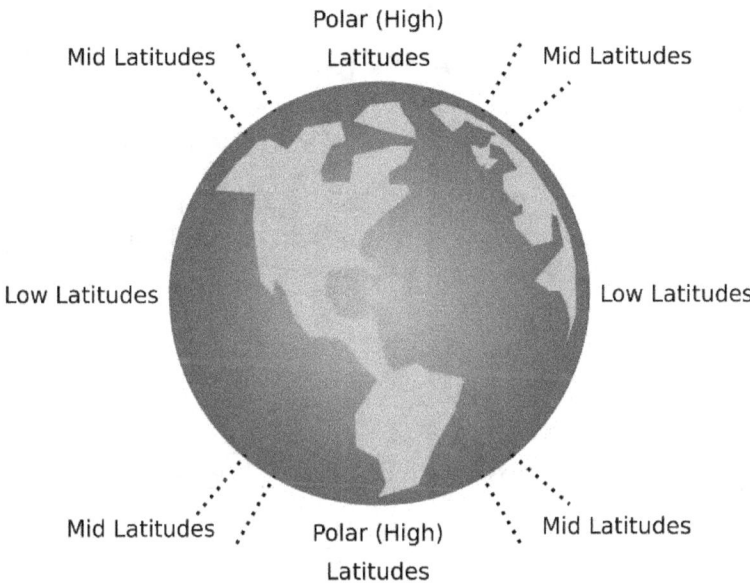

The polar latitudes see by far the most auroral activity while the low latitudes seldom see any auroras. The middle latitudes are usually home to those "ionospheric troughs", regions of the F layer where free electron density is somewhat depleted. The low latitudes are known to produce irregularities in the ionosphere that produce "scintillation" – rapidly fluttering reception.

These differences among the latitudes aren't completely understood, but scientists generally relate them all to the effects of the Earth's magnetic field and the magnetosphere on the ionosphere.

The latitudes also don't quite match up with our usual map latitudes, because they are "geomagnetic latitudes." The geomagnetic poles are offset from the geographic poles by about 13.5 degrees, and they're also offset from the magnetic pole. (Who knew there were so many poles?)

Geomagnetic latitudes are latitudes on a map that uses the geomagnetic poles for the north and south poles.

Phenomena associated with the interaction of the magnetosphere and the ionosphere generally align with the geomagnetic latitudes.

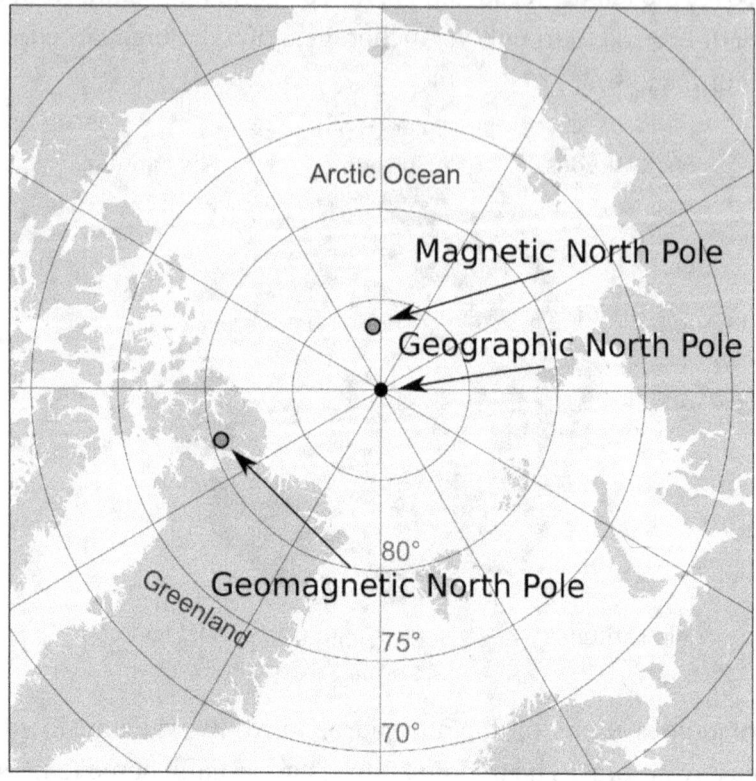

Some very general rules of thumb: The low, or equatorial, latitudes are where you'll most often find the most dependable ionospheric conditions for successful propagation. (Lucky for us continental US hams. Not so lucky for our fellow hams up in Alaska.) The middle latitudes are usually home to the "mid-latitude troughs" – weak areas of the F layer. The polar latitudes can be the land of radio chaos with even moderate geomagnetic disturbances.

Seasonal Variations

The daytime ionosphere also varies by season, because the angle of the sunlight changes. Generally speaking, during the summer months, the ionosphere gets higher and the layers spread vertically, while during the winter months, the ionosphere gets shorter and the layers compress together.

Nighttime variations are much less – the winter night ionosphere is, overall, a carbon copy of the summer night ionosphere. The E layer almost vanishes, and the F1 and F2 merge into a sing, less highly ionized F layer.

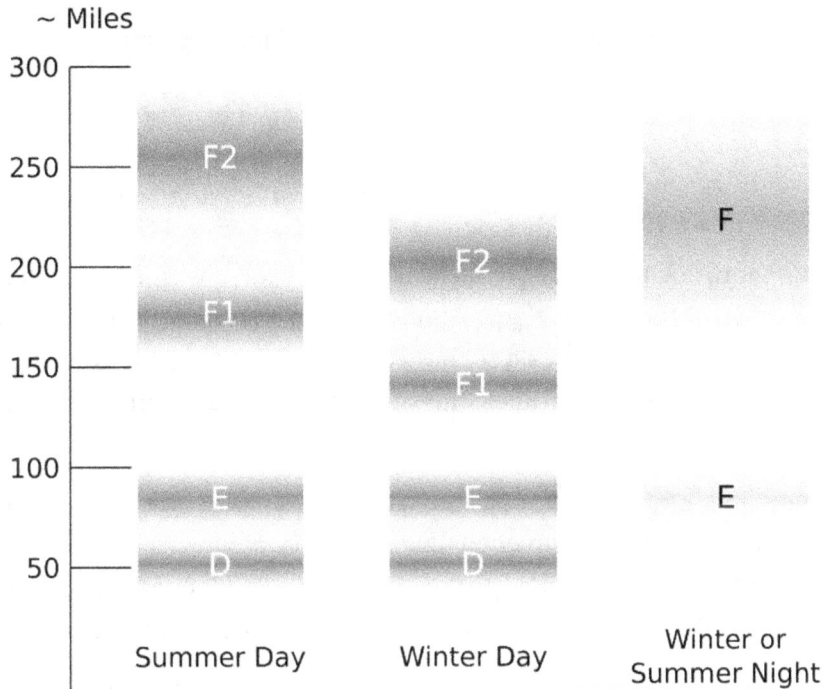

How the Ionosphere Bends our Signals

You had to learn the official definition of a radio wave for your Technician Class License and probably remember that it's *an electromagnetic wave.* You probably also know that radio waves are a lower frequency form of light. Let me ask you a fiendishly simple question: what *is* an electromagnetic wave, *exactly*? When you push the PTT button or hammer that code key, what, precisely, is flying out of your antenna?

I won't hold my breath for your answer, because the answer today is, we still can't answer that question without raising more difficult questions and mind-warping paradoxes. It's a dive down a rabbit hole!

We'll keep it pretty simple, though, without burying you in quantum physics equations, relativity theory, and Great Unanswerable Philosophical Questions. Still -- this gets a little weird. Brace yourself.

Yes, a radio signal absolutely is a wave.

And, a radio signal is absolutely not a wave, it's a bunch of particles.

Put a little more accurately, if we measure a radio wave with instruments that measure waves, it turns out to be absolutely, positively a wave and definitely not particles. BUT! If we measure that very same radio signal with instruments that measure particles, the signal is absolutely particles, no way it could be a wave.

Understand, this isn't a failure of language or vocabulary. It's not a wave OF particles. We don't need to invent a new word for radio waves, like "wavicles" (actually a real physics word) or "partaves" (actually not a real physics word -- I wonder why?) because it's really an either/or situation.

Most folks have some grasp of how a digital camera sensor works. Basically, it's a flat plate with millions of tiny little light sensing elements called pixels. The electrical output of each pixel changes in proportion to the light that falls on it. We measure the output of each of the cells, put it all together with a computer, and, ta daa! We have successfully deployed billions of dollars' worth of research and development to produce a picture of a cat.

Let's say we have a digital camera and we put a dark piece of glass in front of the lens. If light is a wave, what we should see is just lower output from each pixel; but we don't! What happens is *fewer pixels light up*. Each individual pixel that lights at all "lights up" just as brightly as before, but there are fewer "lit up." So obviously, light and, thus, a radio signal is made of particles, and we even have a name for these particles; they're *photons*, so you know they must be real.

But wait, there's more! We know from our VHF/UHF experience about multipath interference, right? Multipath happens when a signal gets bounced around and takes several paths of different lengths to our receiver, some of the signals arrive out of phase, the peaks and valleys interfere with each other, and we end up with lousy reception.

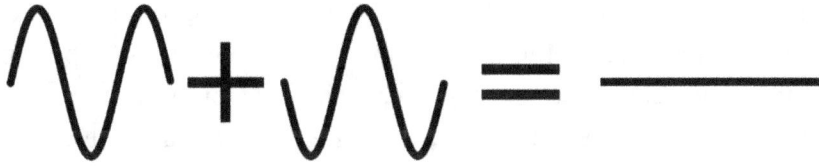

Wait. Wait just one darn minute, hold the phone!

Peaks? Valleys? Phase relationships? What nonsense is this? Particles don't have those, they're just discrete little packets of energy! You don't add stuff to stuff and end up with no stuff! What do particles care if another particle does or doesn't arrive at a particular time? Imagine a thousand ping pong balls launched in a stream at a target -- and just in case things get complicated, imagine these are magic ping pong balls that have no mass and just pass through each other with no collisions, since that's what photons do. How would one ping pong ball know that another ping pong ball had taken a slightly different path? That couldn't possibly make any difference. Proof positive, radio signals and light are waves. By the way, the same sort of demonstration can be done with light, but hey, we're radio people.

This dilemma first started to rear its ugly head back in the 1600's with that fellow we met earlier, Christiaan Huygens. Remember, he figured light was a wave. Another smart guy, Isaac Newton, figured it was "corpuscular" – it was particles. In fact, Newton was such a Big Deal that Huygens' theory gained little traction. No one anticipated that they'd both turn out to be right.

I warned you this would get weird. It's so weird that we have to fall back on metaphors, analogies, and not-quite-accurate statements to wrap our minds around it. Don't worry, physicists do the same thing.

For our purposes, for the moment, we're going to go with the particle model, and we're going to add in a little fiction which is that these particles have a particular size depending on the frequency of the signal. This is not true. For one thing, photons aren't even really particles, they're "quanta", and they have no mass, and no "size" as we usually think of size, since they're not a "thing" the way we usually think of things. However, this is a very useful bit of science fiction when it comes to understanding this topic enough to be able to think logically about DX'ing, which is, after all, the object of the game here.

Back to the ionosphere. Up there, air molecules are getting blasted by solar radiation. If there's a lot of radiation, a lot of electrons get blasted loose and there's a lot of electrical charge floating around up there. Most importantly, those charged particles are closer together than they would be in times of low solar radiation, *and* they exist at greater densities at lower altitudes. So far as we're concerned, the only useful stuff in the ionosphere is those free electrons – we don't much care about the ions they left behind.

Now imagine a photon from our antenna blasting up to the ionosphere. One of three things is going to happen next.

1. Your signal zaps straight through the ionosphere, never to be heard from again by anyone on Earth;

2. Your signal gets refracted back to Earth, or;

3. Your signal gets absorbed by the ionosphere.

Let's say we're transmitting on 40 meters. That means, for our little fantasy, that the photon is 40 meters wide. It gets up to the ionosphere and encounters a bunch of free electrons about, say, 39 meters apart from each other. (Again – this is fiction.) BAM! That signal gets bent, at least a little bit, from its straight path to outer space. It encounters a few more free electrons spaced 39 meters apart --- then a few more -- and the next thing you know, it's headed away from the charged particles and back toward Earth, where some ham 1000's of miles away is anxiously listening for your signal. That's a simple version of *skip*, the phenomenon that makes worldwide radio communication possible. Your signal has become what we call a skywave – as opposed to a ground wave. In practice, that skywave might come back to Earth, bounce off the Earth, head back up to the ionosphere, get bounced again, and maybe even repeat that cycle up to five times before reaching that distant antenna.

But what if our 40-meter photon heads up to a quiet ionosphere and finds electrons spaced 80 meters apart? Whoosh --- it probably shoots straight through, and unless you're trying to QSO with the aliens on Planet X, you're out of luck for the night.

More ionosphere activity – more charged particles closer together -- equals higher frequencies will bounce.

Imagine you're holding a beach ball that's two feet in diameter. You're trying to bounce it off a picket fence, perhaps while blindfolded. If the fence's pickets are an inch apart, that ball will bounce back every time.

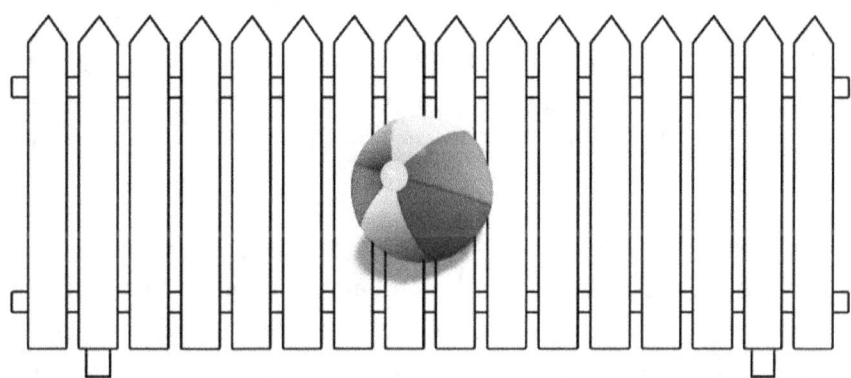

If they're three feet apart, it won't bounce back nearly as often. You'd need a four foot beach ball (or a lower frequency radio wave) to make that work dependably.

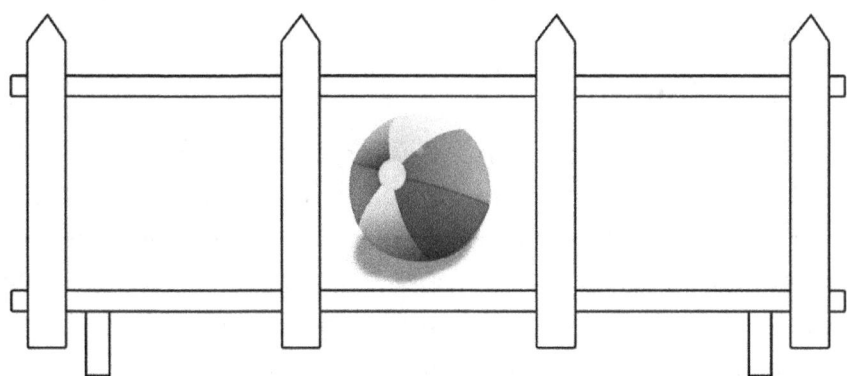

So 160 meters -- our lowest common ham radio frequency, the one with a colossal wavelength, the "biggest beachball" -- must be the absolute best for skipping signals around the globe, right?

Before I answer that, it's important to note, at this point, that when we bounce a signal off the ionosphere, our signal isn't really bouncing like a ball bouncing off a wall. It's being absorbed and reradiated by those free electrons. The electron is sitting there, our 40-meter signal hits it, and the electron says, hey, extra energy -- it's making me all tingly and weird at a frequency with a wavelength of approximately 40 meters! Then, it releases that energy and our signal goes its merry way. It's like we have moved our transmitting antenna to a spot a couple of hundred miles high. That "antenna" acts as though it is somewhat directional because it is operating in an electric field that deflects radio waves back toward Earth.

So, if a long wavelength is the answer, the 160-meter band must be the best frequency for working skywave, right? Well, no.

Remember, there isn't just a MUF, a Maximum Useable Frequency. There's also an LUF -- the Lowest Usable Frequency. Get below this frequency and the signal just gets absorbed by the lower layers of the ionosphere and never gets reradiated.

Huh? Yes. There's something else about that picket fence: the pickets are little devices of destruction for your beach ball. They're all on a timer that switches their destructive power off and on. The rate of the timer depends on how close together the pickets are; the closer they are, the faster they switch.

Whenever the pickets have their destructive power turned on, your beach ball will magically disappear when it hits them! You need enough pickets close enough together to bounce your beach ball, but not so close together that the fence keeps eating your beach balls.

Ionospheric Absorption

Remember, in order to reradiate your signal, an electron must remain a free electron for at least the duration of one cycle of your signal. If it finds an ion that's missing an electron before it can reradiate the signal it just absorbed, it will join that atom and give up the energy of your signal as heat. Even if it just slams into a neutral atom, it will give up some energy as heat. If that happens, you're not communicating, you're just warming up the ionosphere with your watts.

So imagine the pickets on the fence are constantly flashing in and out of their destructive state. If they're far apart, they last longer; in the upper layers of the ionosphere, free electrons can last hours, even days. If they're close together, such as in the D layer, free electrons might only last a microsecond or less. At a microsecond, any frequency under 1 MHz is too low and will get absorbed. Since the 160 meter band starts at 1.8 MHz – well, any recombination faster than 0.56 (1/1.8 x 10⁶) microseconds is going to eat 100% of our signal. Even at a full microsecond, chances are good our signal got there "too late" for about half of the free electrons, because they've already been out for a half-a-microsecond.

Of course, anything could happen with any individual free electron, but we can calculate the *average* rate of ionization and recombination for a given layer under given conditions. Physicists call that rate of ionization and recombination the "*plasma frequency*." In simple terms, plasma frequency is "how long a free electron is available to reradiate our signal."

Longer waves -- lower frequencies -- tend to be absorbed more than shorter waves -- higher frequencies. In fact, the ratio of absorption of two different frequencies can be expressed mathematically as the inverse square of the frequency. Simply put, if you halve the frequency, you get four times as much absorption.

Let's see how this all stacks up, layer by layer.

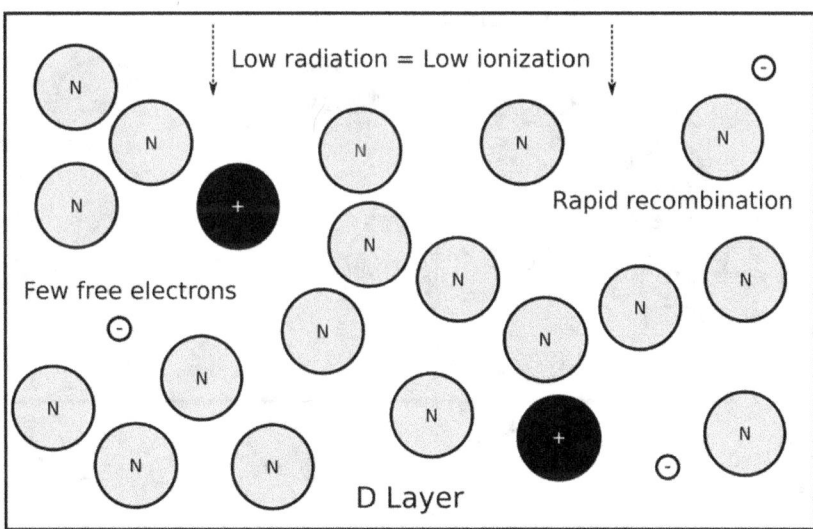

First, the lowest layer, the D layer:

In the D layer we have, relatively speaking, lots and lots of molecules to ionize but not much radiation to do so – most of the radiation gets absorbed in the upper layers. Because the molecules are in close proximity, the few electrons that get free are rapidly re-absorbed.

Since the MUF is determined by the amount of space between free electrons, and the D layer's free electrons have a lot of space between them, "MUF" will be relatively low. Higher frequencies just shoot straight through to space. (As you'll see, in the case of the D layer, "MUF" might as well stand for "Maximum UN-useable Frequency.)

At the same time, because of the proximity of those other molecules, the recombination rate in the D layer is very rapid. That means only higher frequencies could possibly be reradiated – lower frequencies will be absorbed before they can be reradiated. In normal conditions, that means when it comes to propagation, there is no frequency high enough to get through the D layer *and* be refracted back to Earth by the layers above, and there's no frequency low enough to be reradiated that doesn't get absorbed. For us, almost always, the D layer is a dud!

Shortly after sunset, the D layer starts to disappear. Because it is higher, and stays in the sunlight longer, the E layer persists longer than the D layer.

How about the E-layer?

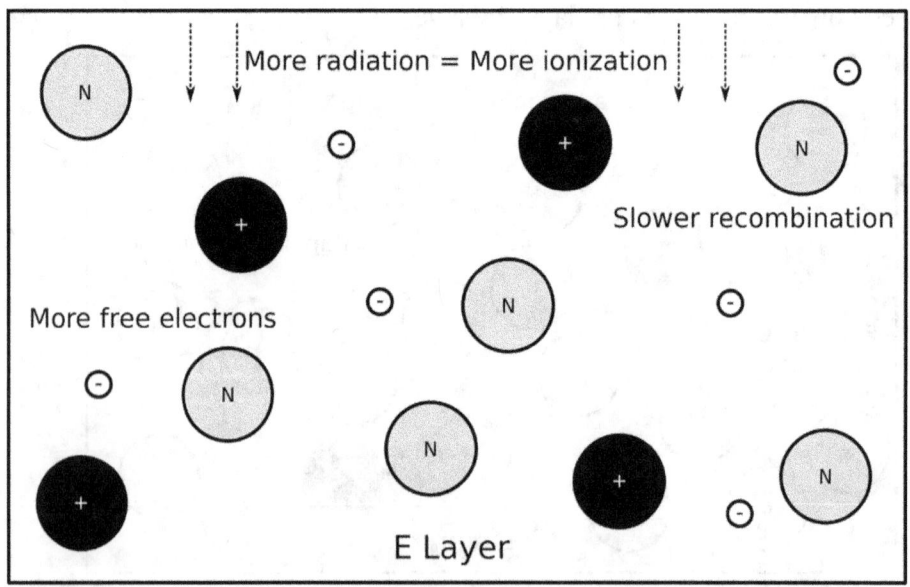

In the E layer, the molecules have thinned out. The radiation level is higher than in the D layer. We have more free electrons, but we also still have relatively rapid recombination happening – though it is slower than in the D layer.

During the day, most of the time, we can't get anything through the D layer that can be refracted by the E layer, and that adds up to zero or very limited propagation from the E layer. However, when isolated pockets of very high free electron density occur in the E layer, we have possibilities for low VHF (10-meter or 6-meter, usually) propagation. Those frequencies can get through the D layer and under these conditions can be refracted by those isolated pockets of high free electron density. That's what we call sporadic E propagation. Lower frequencies won't work in sporadic E because the recombination rate is still too high in the E layer, but in a sporadic E pocket, there are enough free electrons to refract a higher frequency signal.

Finally, there's the F layer, the highest region of them all, up there where there's barely any atmosphere at all.

Now the atmospheric density has really thinned out, and the radiation is at its maximum. Lots and lots of free electrons get created, and they last a long, long time; they last so long that there isn't a significant LUF for this layer, just an MUF, determined by the spacing between the free electrons.

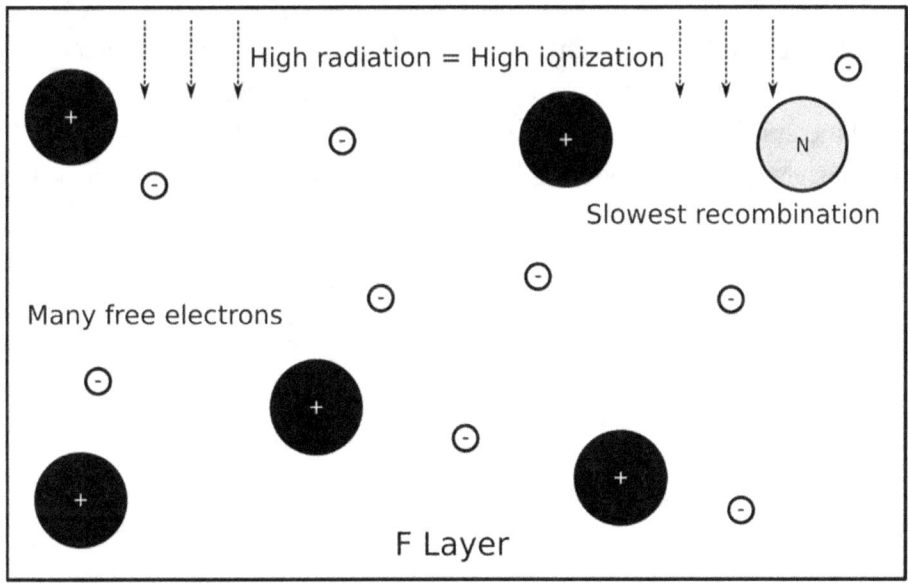

This is why the F layer is, with even a little help from solar conditions, rich, rich territory for propagation. There are plenty of free electrons to refract a wide range of frequencies and the absorption rate is very, very low.

In normal practice, we don't talk about "the MUF of the D layer", we're just talking about the overall MUF – and maybe the LUF -- between two points.

MUF and LUF are somewhat independent, so it's entirely possible to have conditions where the Lowest Useable Frequency is *higher* than the Maximum Useable Frequency which means it's probably a good time to fire up your VHF or UHF rig and chat with some local folks, because HF's going to be pretty useless.

What Frequency?

What's going to be your first choice for a frequency, given the current conditions? What we're looking for is the Goldilocks frequency; a frequency high enough to punch through the layers between us and the F layer, but low enough to *not* punch through the F layer. That frequency turns out to be just below the MUF, the Maximum Usable Frequency, and experienced DX'ers usually recommend a rule-of-thumb of about 80% to 90% of the MUF.

I just checked a website that has a MUF map, and this afternoon, the MUF between here in Washington State and Asia is 35 MHz. Okay, so 80% of that is 28 MHz ....how 'bout that? It's looking like it might be a 10-meter sort of day between here and Asia at the moment.

Between here and Europe, right now, the MUF is 12 MHz, a 25 meter wavelength. If I want to talk with Europe, I need to be thinking no higher than 30 meters, probably more like 40. (That site also tells me 40-meter conditions are generally "poor" at the moment, so I might be out of luck.)

Chapter 6 - Space Weather and the Ionosphere

Sunspots and Sunspot Numbers

We know sunspots have been observed since around 350 BC – Chinese astronomers left us detailed notes in the Imperial records. Some 2,370 years later we still don't completely understand the physics of sunspots, but we do know they seem to be the visible parts of "tubes" of high density magnetism within the sun. Imagine a magnetic rope that's being whirled around in the colossal storm that is the interior of the sun. Every now and then the ends get whirled up to the surface, and we see them as sunspots.

Sunspots are dark, compared to their surroundings, because the strong magnetic field slows the convection process that is bringing heat up from the interior of the Sun, causing a slightly cooler area (3000° C) on the surface compared to the surrounding material (6000° C.)

The sunspots themselves have little effect on propagation. They occur at times of increased solar activity, and coincide with increased amounts of ultraviolet and x-ray radiation.

Times of high sunspot activity are times of improved HF sky-wave propagation because of those higher ultraviolet and x-ray levels. Indeed, a high enough sunspot number can even coincide with enhanced propagation up into the low end of VHF, the 6-meter band.

Sunspot activity is expressed by R, the sunspot number, also known as the Wolf number, but most commonly known in the ham radio world as "the SSN." The number doesn't tell you the number of sunspots, it tells you the relative amount of sunspot activity. The formula is:

$$SSN = k(10g + s)$$

"s" is the number of individual spots seen by the observer, "g" is the number of sunspot groups, and "k" is a constant that varies with the observatory's location and equipment.

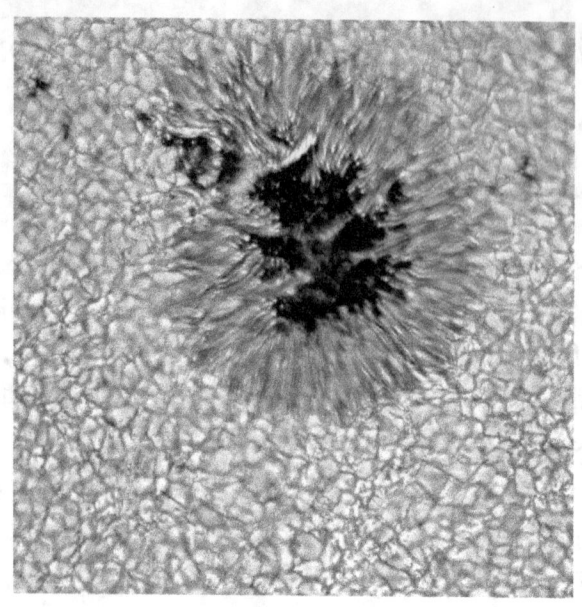
Close-up view of sunspot

As any HF operator who has been trying to work DX for the past few years can tell you, it's quite possible to have a sunspot number as low as 0. In fact, this year, 2018, we went three weeks with a sunspot number of 0 – the longest that has happened in nine years. Historically, the number has rarely gone over 200.

It's also quite possible to have two different yet accurate sunspot numbers for the same day in different information sources. That's because there are, officially, two sunspot numbers! One, The International Sunspot Number, is reported by the Solar Influences Data Analysis Center in Belgium. The other, the NOAA sunspot number, is reported by the US National Oceanic and Atmospheric Administration.

While sunspot activity can and does occur at any time, high sunspot activity is cyclical and tends to peak every 11 years. Nobody knows exactly why the sunspot cycle is **11 years**, so no one knows why sometimes it has been as short as 8 years or as long as almost 14 years (unless that one was really two very short cycles -- the debate goes on.)

At the moment, we seem to be rocketing into the "solar minimum" part of the cycle at a faster pace than normal. The last solar maximum, which hit its peak in 2014, wasn't much of a maximum either.

Another item in the very long list of solar phenomena that are, so far, unexplained is the Maunder Minimum.

From around 1645 until 1715 or so, the Sun seems to have almost totally skipped a few cycles. That coincided with The Little Ice Age. Are those events connected? Here's NASA's way of saying they don't know: "The connection between solar activity and terrestrial climate is an area of on-going research."

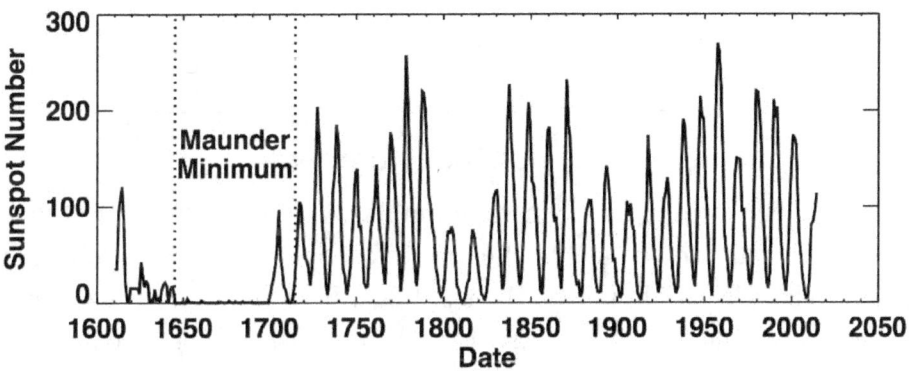

The sunspot cycle is also known as *the solar cycle,* and it's just one of many cyclic events on the Sun that affect propagation.

The Sun is not a homogeneous lump of hot stuff. It's constantly changing, and there are relatively hot spots, relatively cool spots, storms, in addition to those eruptions we call sun spots. As the Sun rotates on its axis, different parts are pointed at us, so the radiation we receive varies. The sun rotates once every 28 days.

You can see we have a cycle within a cycle here. There's a 28 day cycle and an 11 year cycle – and that's just the two major cycles. There are also more random solar events that happen, such as solar flares and solar coronal holes that can create geomagnetic storms here on Earth.

Aside from "conditions should be good," what, specifically, does a high sunspot number mean for us in terms of propagation?

Each layer of the ionosphere as, at any given moment, a particular "critical frequency." That's the highest frequency that will bounce off it when a signal is sent straight up. Higher critical frequencies indicate better propagation conditions. Here's what happens to the critical frequency of various layers during daylight hours as the sunspot number increases:

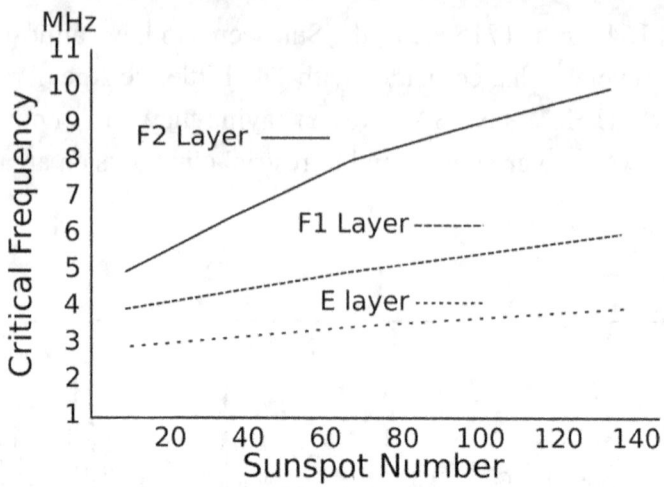

You can see that even dramatic increases in the sunspot numbers affect the D and F1 layers rather moderately, while the effect on the F2 layer is the most dramatic. This plays right into our hands! Regardless of the sunspot number, the D layer is going to almost always disappear at night. The molecules are so close together, the free electrons are going to recombine quickly, and what was a highly absorptive layer turns into plain old air, so it's no longer an impediment to our signals. To a lesser extent, the same process destroys most of the E layer, though it dies away more slowly since the air is thinner. The F layer persists, though. Remember, molecules are few and far between up there, relatively speaking, so the free electrons can stay free for a long time. As the critical frequency goes up, we have a better and better chance of being able to use a signal that is at a high enough frequency to pierce through the remaining E layer while still being low enough to be refracted back to Earth by the F layer. The disproportionate boost to the F layer provided by the increased solar activity indicated by the high sunspot number helps the F layer be more useful without creating a dramatically "thicker" E layer.

The Solar Flux Index

Every day, the Penticton Radio Telescope Observatory, located about 200 miles east of Vancouver, BC, reports the Solar Flux Index. Along with the sunspot number, the SFI is one of our most valuable tools for predicting propagation. The scale, in principle, goes from 62.5 to infinity, but in reality the range is more like 62.5 to 300.

Technically, the SFI is a measure of solar radiation at 10.7 centimeters wavelength. It's reported in Solar Flux Units. Typical solar flux numbers will range from 50 to 300 SFU's. Higher fluxes usually mean a higher MUF, lower numbers a lower MUF, and that overall conditions will not be wonderful. However, it takes some time, like a few days, for a change in solar flux to change conditions in the ionosphere, so Monday's report of a 250 SFI might not make for great DX'ing until Wednesday or Thursday.

There's one SFU value for the whole planet, unlike other values we'll study below.

10.7 centimeters is the wavelength of a 2800 MHz signal. Why do we measure at 10.7 centimeters? Why not a nice even 10, for instance? I've looked all over for this answer! It seems when Arthur Covington, the Canadian astronomer who first measured solar flux in 1946, did his first observations he used 10.7 centimeters and that established the *de facto* standard. And that was that, since changing the standard would make all the data collected since then useless for comparison purposes.

Solar Flares and Sudden Ionospheric Disturbances

To sum up the previous section: Sunspots good.

Solar flares, and the stuff they cause, like Sudden Ionospheric Disturbances and geomagnetic storms; not so good, except for VHF/UHF aurora bounce.

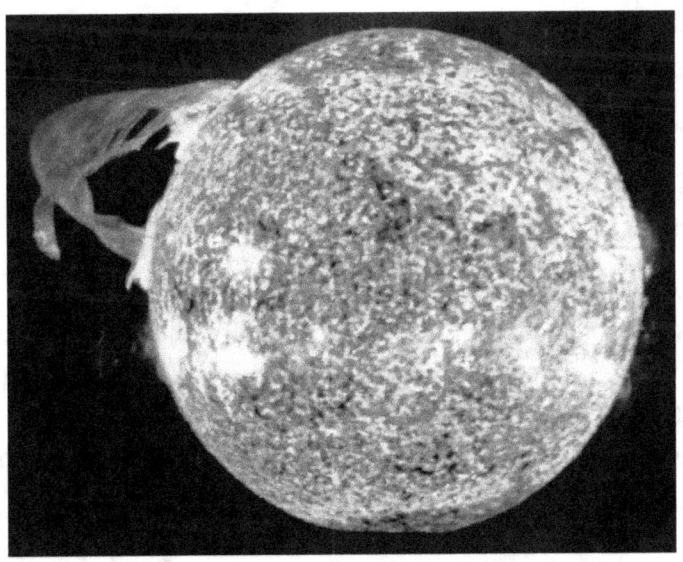

Solar Flare

Sudden Ionospheric Disturbances are caused by solar flares. A solar flare is a stupendous release of energy caused by the sudden collapse of a magnetic line of force surrounding the sun. The picture on the preceding page is a spectacular one seen from Skylab.

The ends of that "arch" in the solar flare picture are about 365,000 miles apart. You could fit the Earth between those ends 30 times. Some solar flares release energy equivalent to 100,000,000,000 tons of TNT being exploded *every second*. 100 *billion* tons of TNT – that's a TNT bomb roughly as big as a good sized mountain going off every second! Put another way, it's like someone switched on a 100,000,000,000,000 MW transmitter putting out noise with several billion Hertz worth of bandwidth. The average solar flare lasts about an hour, but the disturbances they cause here on Earth can go on for days.

Of course, when magnetic fields change, electrical fields change as well, and when the enormous burst of electromagnetic energy and solar wind from a solar flare hits our magnetosphere and ionosphere, things change rapidly up there. All the ionosphere layers are enhanced, including the already very absorptive D layer. HF signals disappear – sometimes even atmospheric noise disappears.

Usually, an increase in radiation from the sun is good news for propagation, but a Sudden Ionospheric Disturbance is a classic case of "too much of a good thing" that disrupts our signals and SID's disrupt signals on lower frequencies more than those on higher frequencies.

Where the solar flare is "aimed" makes a big difference in the effects here on Earth because the radiation from them is somewhat directional.

Coronal Mass Ejections

The solar corona is the sun's atmosphere. From time to time, a solar flare knocks a "hole" in the corona, the outer part of the Sun's atmosphere, allowing a massive stream of particles -- also known as the solar wind -- to escape. That's known as a coronal mass ejection, also known as a "solar coronal hole." Notice, now we're not talking about electromagnetic waves, we're talking about actual, highly energetic, fast-moving particles. If the hole is oriented toward Earth, those particles come blazing into our magnetosphere and create a geomagnetic storm.

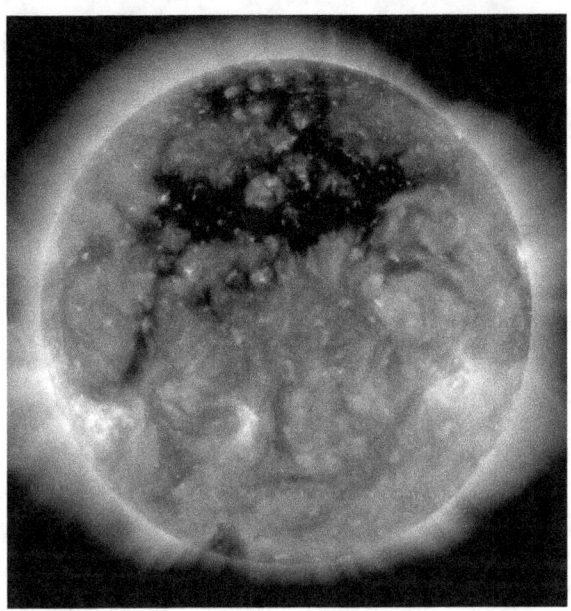

The dark area is a big coronal hole

In a geomagnetic storm, HF communications are disturbed, and most likely "disturbed" to the point of being destroyed. For that matter, even the 1.5 GHz signals coming down from GPS satellites are disturbed. We'll also probably see auroras extending down from their usual home in the polar latitudes, sometimes far down into the middle latitudes.

While the *light* from the Sun takes 8 minutes to get here, and nothing travels faster than the speed of light, the charged particles from coronal mass ejections aren't traveling at the speed of light. By the time they get here *and* the ionosphere reacts to the events they cause in the magnetosphere, 20 to 40 hours have gone by, so we can get a lot of warning about these coming disturbances to HF propagation.

Geomagnetic Storms

When a solar flare rips a hole in the Sun's corona, and that hole is aimed at us, the result for Earth is a geomagnetic storm.

The Sun is constantly pushing photons and other particles out into space. This constant flow is called the solar wind. In propagation science, it's called the solar flux, and you'll see that term when you start looking at space weather web sites.

If you've ever been in a strong wind, you know that from time to time a really strong gust will come along.

It's the same way with the solar wind. From time to time, some event will occur on the Sun that creates a "gust" in the solar wind, such as a big hole in the corona.

Remember that picture of the magnetosphere? When that gust of solar wind gets to Earth, it stretches or compresses different parts of the magnetosphere and increases the magnetic energy present. As we know, moving lines of magnetic force create electrical voltages and vice versa, so when that magnetosphere gets squeezed and jostled all sorts of interesting phenomena start to occur, and that set of phenomena is what we call a *geomagnetic storm*. The most visible of these phenomena is an increase in auroral activity.

One effect of that gust of solar wind is to open up passages into Earth's polar regions for the fast-moving particles it carries. Now we don't just have ionization occurring in the ionosphere, we have raw, free electrons being pumped in at colossal rates, and the skies light up.

Auroras -- the Northern Lights -- are usually only seen in the higher latitudes, near the poles. A powerful geomagnetic storm can create auroras much closer to the equator. In 1989 a particularly powerful geomagnetic storm had auroral activity visible as far south as Texas.

HF's dead, Jim – time to bounce some VHF off those pretty auroras!

The Planetary K-index

In addition to the Solar Flux Index, we can look at the Planetary K-index and the Planetary A-index. These indices tell us about the space weather up in the magnetosphere. Is it a calm day up there, or is there a storm going on?

The K-index is measured at various magnetic observatories, so different places report different values. The K-index is, basically, a three-hour snapshot of the state of the magnetic field, so the K-index is a measure of the short term stability of the Earth's magnetic field.

K-index values are *quasi-logarithmic*. They'd be regular old logarithmic except for the fact that they're measured at multiple locations around the world at different times of day throws off the math a little. They're (almost) like decibel values, where 3 dB is a doubling of power, 10 dB is 10 times power, etc. K-index values from 0 to 1 indicate a quiet day in the magnetic field. Values of 1 to 5 indicate a minor or moderate solar storm. Values over 5 indicate major storms that severely hamper or black out HF altogether. A 9 would be a genuinely scary event that would wipe out all sorts of electronic devices, not to mention playing havoc with the electrical transmission system. (A value of 10 is the apocalypse. Take appropriate steps.)

All the daily K-index values are then manipulated mathematically to create the daily A-index.

Imagine a bar magnet under a piece of paper with the iron filings sprinkled to show the magnetic lines of force.

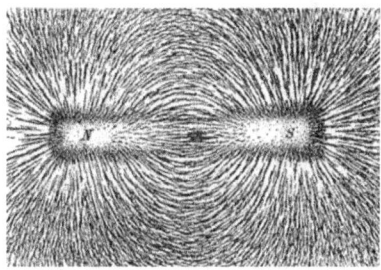

The lines of force converge near the magnet's poles. So do the magnetic lines of force around the Earth, and that affects the behavior of the ionosphere at the poles. For our purposes, this tends to make transpolar propagation – sending signals over the *polar paths* – more problematic because the ionosphere is often more absorptive in those areas. That's especially true when the A or K index is elevated, and that's when we're also likely to see more in the way of aurora activity – not a coincidence!

The A-index

The K-index is the short term stability of the geomagnetic field. The A-index is a measure of the long term stability of the Earth's geomagnetic field. In this case "long term" means "one day." Because of the complex nature of the mathematics applied to create the index, the A-index runs from 0 to 400.

Once the K-index and A-index values are calculated for all the various magnetic observatories, a pair of global indicators are generated, known as Kp and Ap, for "planetary" K and A.

High K (and hence, A) values work *against* high solar flux levels when it comes to radio. We like high solar fluxes with low K's and A's. High K's and

A's mean the Maximum Usable Frequency's coming down, maybe even below the Lowest Usable Frequency.

Other Space Weather Indices

Electron Flux: The solar wind consists mostly of a form of matter called plasma. Plasma is made up of negatively charged free electrons and positively charged ions – and if that sounds familiar, it's because of the ionosphere's various percentages of plasma. One value NOAA reports daily is the Electron Flux index. This is the density of free electrons in the solar wind. The scale runs from zero to "we don't actually know how high this can go." 1000 is a *big* number, though, and is associated with near total HF blackouts through the polar latitudes. Under 100 has little to no impact on HF. If you're planning some trans-polar DX'ing, this is a good number to keep an eye on.

Proton Flux: The density of protons in the solar wind. Like Electron Flux, the scale goes from 0 to ???. Both proton flux and electron flux most influence the E layer.

Solar Wind: The SW index tells you the speed of the solar wind in kilometers per second. The scale goes up to 1000, but that's extreme. Values over 200 km/sec have some impact on HF communication, with values over 600 having fairly severe impact.

X-ray Flux: The GOES (Geostationary Operating Environment Satellites) hover in geostationary orbit over the Earth, mostly keeping track of weather events down here on Earth. However, they also keep an eye on space weather. Among the things they track is the X-ray Flux index. High X-ray fluxes indicate solar flares, solar coronal holes or even coronal mass ejections – big explosions in the sun's corona which can launch massive amounts of particles. As you might imagine, if those particles hit the magnetosphere, it can affect the ionosphere; usually in a negative way from our point of view.

X-ray flux values have an odd scale that matches up with solar flare values. They run from A 0.0 through A 9.9, then B 0.0 through 9.9, C 0.0 through 9.9, then skip to M 0.0 through 9.9, and X 0.0 through, supposedly, 9.9. B's are 10 times more powerful than A's, and each letter represents another 10x increase. That puts the X level at 10,000 times the A level. At the X level, each increase of 1.0 is a doubling of the original. Any X is a *big* solar flare – an X 9 is about 5,000,000 times an A.

I said "supposedly" about the X 9.9 value because in 2003 a solar flare was recorded that blew past the satellite's ability to measure it at X 28. They think it might have hit X 45.

304A: The 304A solar parameter is the measure of ultraviolet emissions at 304 angstroms, correlated to the solar flux index.

The solar flux index was standardized back in the 1950's – it's a measure of the sun's radiation at the frequency of 2800 MHz, a wavelength of 10.7 cm. Think of it as a broad view of the Sun's activity. Very useful, and now we have daily historical data going back some 60 years for comparison purposes, so the solar flux index isn't going away. However, it was created before we had satellites. We don't get much of the Sun's UV radiation down here on the surface, and the amount varies wildly with atmospheric conditions. But stick a satellite up above the atmosphere, and we can get a super-accurate reading of the activity in the UV parts of the spectrum, and that turns out to be a more accurate and immediately useful measure than the solar flux index.

The "A" in 304 A stands for angstrom units, equal to one $10,000,000^{th}$ of a millimeter. 304 angstroms is a wavelength of ultraviolet light.

304 A is responsible for the majority of the ionization of the F layer.

B_z: B_z indicates the direction and strength of the interplanetary magnetic field – the magnetic field between us and the Sun. The interplanetary magnetic field is mostly the Sun's magnetic field. It gets measured and quantified in several ways. Among those ways are the B indices – B_X, B_Y, and B_z. B_X measures the orientation and strength of the field on the X axis of an imaginary three-dimensional grid, with the X axis defined by the plane of the ecliptic – the plane traced by the Earth's orbit. In plain terms, "left or right as we stand at the North Pole and look at the Sun." B_Y tells us about the Y axis – toward the Sun or away from it -- and B_Z – the one with the most effect on our own magnetosphere's events – shows the direction and strength of the interplanetary magnetic field on the Z axis. In other words, B_Z tells us the direction of the North/South axis of the Sun's magnetic field relative to our own field and how strong it is.

The interplanetary magnetic field is normally aligned the same as ours – pointing northward. Unlike Earth's relatively stable field, though, the Sun's can change orientation radically, especially when there is a lot of sunspot activity.

If you hold a pair of bar magnets so the north poles and south poles are facing each other, they attract. If you reverse them, they repel. If you had your piece of paper and jar of iron filings, you'd see the combined lines of magnetic force of those magnets shift dramatically as you reversed the magnets.

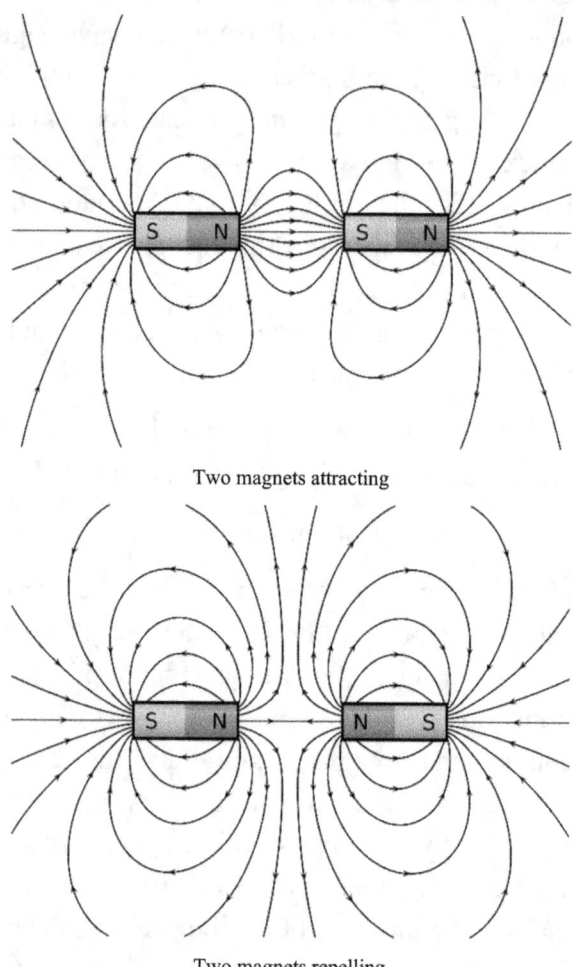

Two magnets attracting

Two magnets repelling

The same thing happens on a gigantic scale with the Sun's magnetic field's interaction with the Earth's.

We say the Earth's magnetic field points northward, so when the Sun's magnetic field points **southward**, the magnets are attracting. The effect is solar wind particles can more easily enter the upper atmosphere, especially at the poles, and interact with the ionosphere to affect propagation, whether for good or ill. Usually ill.

Space weather information is easily available from multiple internet sources. By all means, check out physicist "Dr. T" Tamitha Skov's Solar Storm Forecast on YouTube, for one.

For all the info put together in "ham friendly" style, as well as lots and lots of information about space weather and propagation in general, you must visit this site put together by Paul Herrman, N0BNH.

http: www.hamqsl.com.

Paul has created a "widget" for web sites that gives a summary of all the latest solar-terrestrial data and even includes a general prediction of propagation conditions. The data is also available as a "gadget" for Windows. When you go to the site, you'll find something like this:

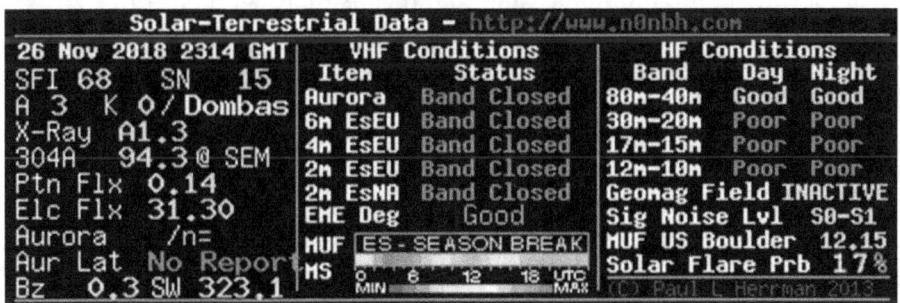

Obviously, that's written in shorthand. Here's a guide:

- SFI is Solar Flux Index.

- SN is the sunspot number.

- A and K are the Planetary A and Planetary K indices.

- X-Ray is the X-ray flux index, the one that correlates to solar flares.

- 304A is the "304 angstroms" measurement. The "@ SEM" tells you it is the measurement of 304A from the Solar EUV Monitor on the Solar and Heliospheric Observatory satellite.

- Pf is the Proton Flux, the density of charged protons in the solar wind. Pf mostly affects the E layer.

- Ef is the Electron Flux – the density of electrons in the solar wind. Ef also affects, mostly, the E layer.

- The Aurora listing is an estimate of how low a latitude the Northern aurora will reach. That particular listing in the illustration says "no aurora activity."

- Bz is that one that tells us the orientation of the Sun's magnetic field relative to Earth's, and finally,

- SW is the speed of the solar wind in km/s.

- The HF conditions listed are predictions and, obviously, are very general.

- The VHF conditions box is all about aurora and sporadic E, listed as E_S. In the illustration, there is no latitude for the aurora because it's a quiet time at the moment – and that means there's no Aurora propagation, which is the next thing listed. Then you see the sporadic E conditions for the various VHF bands.

- EME is Earth-Moon-Earth conditions.

- You can see today we have a 19% probability of a solar flare.

- The MUF listed in that part of the widget is for sporadic E.

- MS is Meteor Scatter, which is looking pretty strong today.

- The geomagnetic field is Very Quiet.

- The Sig Noise Lvl is the Signal Noise Level, not to be confused with a signal *to* noise *ratio*. It's an estimate of how much noise is being generated by the interaction of the solar wind with the magnetosphere.

- Finally, you see the MUF as reported by NOAA – this particular one is for Boulder, CO.

Now, what does one do with all that information? It does look a bit daunting, but happily, Paul Herrman also created a brilliant chart that makes it all make a lot more sense. His chart is reproduced here, with his kind permission.

Understanding HF & VHF propagation conditions using data from N0NBH's HAMQSL Solar Data Panel

Category	Radio Blackouts	Solar Radiation Storms	Geomagnetic Storms	Band Openings	Electron Alert
Look At:	X-Ray	Proton Flux	K-Index/K-nT/Aurora/Solar Wind/Bz	Solar Flux	Electron Flux
Extreme	X20 — Complete HF blackout on entire sunlit side of Earth lasting hours.	1,000,000 — Complete HF blackout in polar regions.	K=9 (nT=>500) [Aur=10++] (SW=>800) [Bz=-40 -50] HF impossible. Aurora to 40°. Noise S30-.		
Severe	X10 — HF blackout on most of sunlit side for 1 to 2 hours.	100,000 — Partial HF blackout in polar regions.	K=8 (nT=330-500) [Aur=10+] (SW=700-800) [Bz=-30 -40] (100 per cycle) HF sporadic. Aurora to 45°. Noise S20-S30.	200-300 (SN=160-250) Reliable communications all bands up through 6m	>1000 Alert Partial to complete HF blackout in polar regions
Strong	X1 — Wide area HF blackout for about an hour on sunlit side.	10,000 — Degraded HF in polar regions.	K=7 (nT=200-330) [Aur=10] (SW=600-700) [Bz=-20 -30] HF intermittent. Aurora to 50°. Noise S9-S20.	150-200 (SN=105-160) Excellent conditions all bands up through 10m w/6m openings	
Moderate	M5 — Limited HF blackout on sunlit side for tens of minutes.	1,000 — Small effects on HF in polar regions.	K=6 (nT=120-200) [Aur=9] (SW=500-600) [Bz=-10 -20] HF fade higher lats. Aurora to 55°. Noise S6-S9.	120-150 (SN=70-105) Fair to good conditions all bands up through 10m	<1000 Active Degraded HF propagation in polar regions

Minor	M1 Occasional loss of radio contact on sunlit side.	100 Minor impacts on HF in polar regions.	K=5 (nT=70-120) [Aur=8] (SW=400-500) [Bz=0 -10] HF fade higher lats. Aurora to 56°. Noise S4-S6.	90-120 (SN=35-70) Fair conditions all bands up through 15m	<100 Active Minor impacts on HF in polar regions
Active	C1 Low absorption of HF signals.	10 Very minor impacts on HF in polar regions.	K=3-4 (nT=20-70) [Aur=6-7] (SW=200-400) [Bz=0-+50] Unsettled/Active Minor HF fade higher lats. Aurora 60-58°. Noise S2-S3.	70-90 (SN=10-35) Poor to fair conditions all bands up through 20m	<10 Normal No impacts on HF
Normal	A1 – B9 No/Small flare No or very minor impact to HF signals.	No impact on HF	K=0-2 (nT=0-20) [Aur=<5] (SW=200-400) [Bz=0-+50] Inactive/Quiet No impacts on HF. Aurora 67-62°. Noise S0-S2.	64-70 (SN=0-10) Bands above 40m unusable	<1 Normal No impacts on HF

VHF Conditions

Aur Lat (Auroral Latitude): Indicates lowest latitude from the current Aurora Activity measurement. Text color coded for low activity, hi-latitude, & mid-latitude.
Aurora (Northern Auroral Activity): Band Closed = No/Low Auroral activity. High LAT AUR = Auroral activity >60°N. MID LAT AUR = Auroral activity 60° to 30°N.
EsEU (Sporadic E - Europe): Band Closed = No Sporadic E (ES) activity. High MUF (2M only) = Cond support 2M ES. 50/70/144MHz ES = Respective band open
EsNA (Sporadic E - North America): Band Closed = No Sporadic E activity. High MUF = Cond support 2M ES. 144MHz ES = Sporadic E reported >140 MHz.
EME (Earth-Moon-Earth): Current EME degradation. Very Poor (>5.5dB), Poor (4dB), Moderate (2.5dB), Good (1.5dB), Very Good (1dB), Excellent (<1dB).
Solar Flare Probability: Provides the probability of a solar flare (in %) for the next 24 hours.
MUF (Max Usable Frequency Bar Color): No Sporadic E (ES) activity / ES reported @ 6M / ES reported @ 4M / Cond support 2M ES / ES reported @ 2M
MS (Meteor Scatter) Activity Color bar: Provides meteor activity color coded MIN to MAX conditions (see the graph below the bar).

©N0NBH Paul L Herrman 2010

.Chapter 7 -- Types of Ionospheric Propagation

Single Hop Skip

In the simplest form of ionospheric propagation, our signal departs our antenna, travels up to the ionosphere, gets bent back to Earth, and is received – and that's pretty much the whole story. One trip up, one trip down, the end.

Single Hop Skip

It would not be unusual to make single-hop contacts during the day or night. Clearly the distance of the hop will depend a lot on the takeoff angle of the signal and, of course, on favorable skip conditions.

Given the geometry of the situation, there's clearly a maximum distance for single-hop skip, depending on which layer is doing the refraction. For the E layer, it's around 1,000 miles, for F2, up to 3,000 miles.

While we're talking about skip, we should clear up some terminology. While we talk about "working skip" when we successfully use a skywave, the original term, skip, had to do with the "skip zone." The skip zone is the area between the transmitter and the first (or only) touchdown of the signal. (More precisely, it's the area between the end of the ground wave and that first touchdown, but most of our ground waves don't go very far.) In the skip zone, your signal cannot be heard. The signal skips over part of the path. This explains why sometimes you can be talking with someone hundreds of miles away on, say, 10 meters, but your friend across town can't hear you.

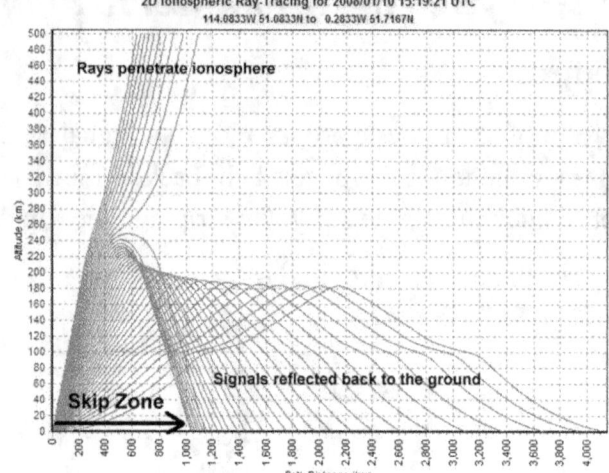

By the way, the plot above is a real propagation prediction done on some remarkable propagation prediction software called PropLab, about which more later.

Multiple Hop Skip

As you might guess, multiple hop skip is like single hop skip with more than one hop.

In multiple hop skip the signal is reflecting off the Earth as well as the ionosphere.

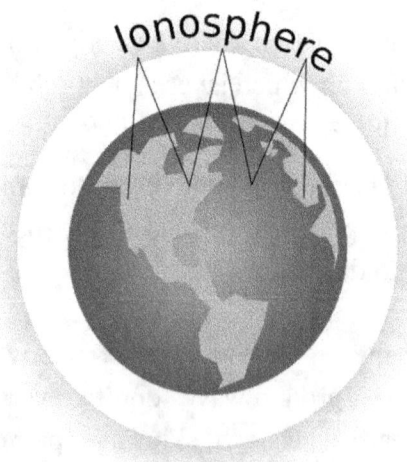

Multiple Hop Skip

Clearly, this requires conditions superior to those that will generate a single hop, and those conditions need to be favorable for all the skip points in the path. Multiple hop happens mostly at night, and if you have a beam antenna, your best chances will be found in the direction of the darkness. If it's early morning where you are, shoot in a westerly direction. If it's evening, head east. Those ground reflections will be a lot less lossy if they're occurring off of water than land, so trans-oceanic is a good way to think.

When the sunspot number smiles on us, it's possible to have five, or even more, skips.

Chordal Hop Skip

If we manage to get just the right takeoff angle, with the right geometric relationship, with the right conditions at the right time of day (whew…) we might create some chordal hop skip. Chordal hop signals cover vast distances with very little loss.

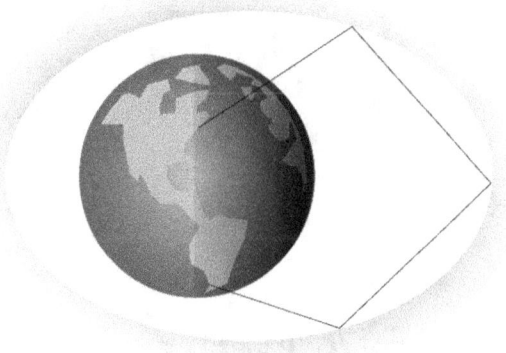

Chordal Hop Propagation

"Chord" in this case refers to the geometry term, not the musical one. A chord of a circle is a line that intersects any two points on the circle. To illustrate this type of propagation so it's understandable, we've stretched out our ionosphere to more closely resemble its real shape.

Because we've caught the right conditions at the right angle, our signal goes a *long* way before returning to Earth. Chordal hops are what can create high-quality *long path* signals, where our signal is going "the wrong way" around the Earth to get to its destination. Obviously, this is a strictly nighttime phenomenon, and a highly desirable one at that.

There's nothing magic happening on a chordal path. The desirability of the chordal path is mostly a matter of distance. The path a chordal hop signal takes is much, much shorter than the path of a normal skip signal. A chordal hop bypasses all that going back down to Earth and going back up to the ionosphere distance, and the inverse square law tells us that every time we cut the distance our signal travels in half, it arrives four times more powerful. Chordal hops also cut out loss from ground reflections.

Ducted Hops

On the sunny side of the world, it's sometimes possible to propagate long distances, especially on the longer wavelength bands, via ducted hops.

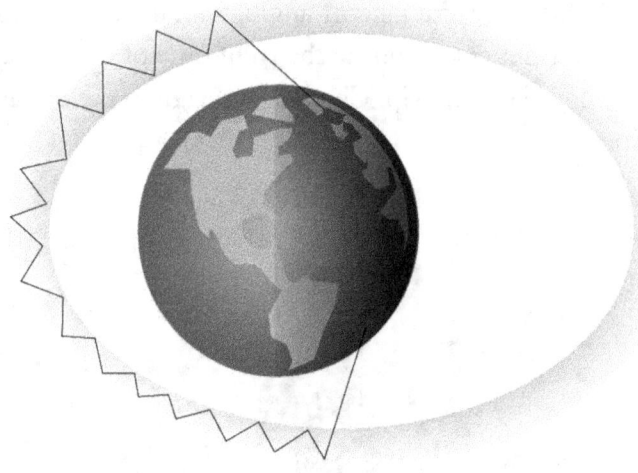

Ducted Hop

These occur when the signal gets trapped between the F1 or F2 layer above and the E layer below. It's the HF version of tropospheric ducting.

Like most of the ionosphere game, this one is mostly a matter of catching a good angle on some good conditions. It helps to be lucky!

Ducted hops are primarily a 160 meter – known as Topband to its fans – phenomenon. If you enjoy propagation puzzles, get into 160 meter – there's a *lot* that isn't known about how 160 meter propagates.

Gray Line Propagation

Right before local sunrise and after sunset, we're out from under the signal-sucking D layer, but still under highly ionized E and F layers. That borderline between night and day is called the **gray-line**, and it can support very strong skip propagation along that line.

Gray Line Propagation

It may be easier to understand the power of the gray line if you see it on a flat map.

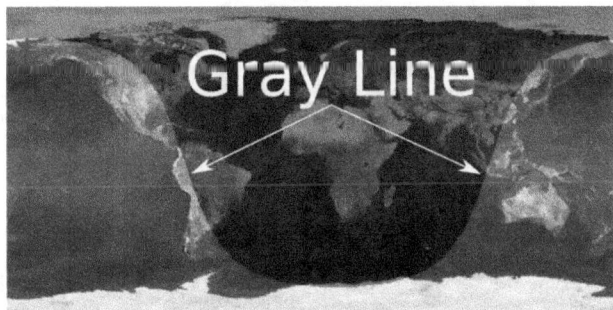

Notice that at the moment this map was made, the gray line could have been providing propagation from Indiana to Indonesia.

Because the Earth is tilted on its axis, the direction of the gray line changes with the seasons. On the equinoxes, it runs directly north and south; at other times it runs up to 23 degrees east or west of due north or south.

Gray line is most often a 10-meter, through 30-meter band propagator – and do not dawdle with your QSO's! The gray line is all about the ionosphere's transition from "day mode" to "night mode" and the transition happens in minutes.

Trans-Equatorial Propagation

Transequatorial propagation is a special form of propagation that occurs between two mid-latitude points at approximately the same distance north and south of the magnetic equator.

The magnetic equator is not quite the same as the usual equator with which we are all familiar, because the magnetic north and south poles are not located at the same spots as the geographic north and south poles. The magnetic equator is offset from the geographic equator. Roughly speaking, it is south of the geographic equator in the western hemisphere, and north of it in the eastern hemisphere, but since we're usually in the US, we're usually only concerned with the western hemisphere.

Transequatorial Propagation

Transequatorial propagation, often known as TEP, is still not fully understood, but we know it most commonly occurs in the late afternoon and evening, and works best in times of high sunspot activity.

Though in the picture it looks a lot like gray line propagation, late afternoon/early evening transequatorial propagation is most often useful on the 6-meter band. Late evening TEP tends to work best on 2 meters and 70 cm. It also is mostly limited to stations that are equal distances north and south of that magnetic equator, and that are within about 15 degrees of each other relative to a line drawn between the north and south magnetic poles. This certainly points to the possibility that it relates to the Earth's magnetic field.

TEP can, under very good conditions, continue until around 11:00 PM, but that's unusual. TEP works best in the afternoon or early evening.

The range of afternoon TEP is from 3000 to 4000 miles, evening TEP is from 2000 to 5000 miles.

Sporadic E

From time to time, for reasons dimly understood for the moment, clouds of high ionization form in the E layer. In these times, propagation of upper HF frequencies and VHF are enhanced. In the temperate latitudes, this can occur during the day or, less often, at night. In the tropics, it is almost exclusively a daytime event.

The curious nature of sporadic E – pools of free electrons that can be as much as 100 times as dense as surrounding areas – is rather unique among ionospheric phenomena. How can one area get more ionized than another when both are getting, essentially, the same amount of the radiation that creates the E layer? Until we fully understand the mechanism(s) behind that, we're going to have a tough time developing a sporadic E forecasting model.

To begin to understand what *is* understandable about sporadic E, first understand that sporadic E is not a single phenomenon.

Normal E layer ionization forms a fairly homogeneous "layer" – a steep increase in free electron density – that is about five miles thick during the day. On the daytime side, the E layer is constantly regenerated by ultraviolet and x-ray radiation, and because the air is dense at that level, it just as quickly breaks down as the ions and free electrons recombine. On the nighttime side, there's less ultraviolet and x-ray radiation, so the recombination process "wins the race" and the E layer disappears shortly after sunset only to reappear shortly after sunrise. That "normal" E layer can't refract signals much above 22 MHz.

When we talk about sporadic E (also known as E_S in scientific literature), we're talking about "lumps" of high free electron density in that otherwise homogeneous layer.

Two types of sporadic E form in the auroral zones.

Most of our atmosphere is not invaded by the plasma that is the solar wind, but the poles are an exception. Because of the shape of the Earth's magnetic field, free electrons come streaming into the poles in greater or lesser numbers, depending on the strength of the solar wind – which is reflected in the K index.

That stream of free electrons can create "night E_S" over the poles. The layer created acts a lot like the daytime E layer, refracting signals up to about 20 MHZ; just barely higher than the "normal" E layer. The only difference is this E layer is present at night.

Very intense solar winds can blow in enough free electrons to create a distinctly different area of E_S, called "auroral E_S." Auroral E_S is much richer in free electrons and can refract signals up to at least 144 MHz. Radar can detect returns off such auroral E_S patches up to 200 MHz.

Both types of polar E_S tend to be "patchy" with varying levels of free electron density through the area.

A similar type of E_S forms in the equatorial latitudes. Typically, this band of equatorial E_S is 100 to 200 miles wide and is near the magnetic equator. It's capable of scattering signals up to 50 or 60 MHz, but is of little use to global communications because it is so narrow.

The mechanisms that create polar and equatorial E_S are reasonably well understood. Perhaps surprisingly, they are related processes, both driven by the Earth's magnetic lines of force. In the case of the poles, the lines of force are nearly vertical, while in the equatorial regions, the lines of force are nearly horizontal (relative to the Earth's surface.) In both cases, the net result is a compression of the free electrons into a thin but dense layer. In the mid-latitudes, though, the slanted lines of magnetic force break up that tendency.

Understanding some of the current thinking about mid-latitude sporadic E requires wrapping our minds around a bit of plasma physics. Plasma is that fourth form of matter, neither solid, liquid, nor gas, but a gas-like substance consisting of ionized atoms. In other words, the stuff the ionosphere is made of, to a greater or lesser extent.

Down here on Earth, when the wind blows out of the North, the wind blows out of the North and that's pretty much that. If a leaf gets caught in the breeze, it goes in the direction the breeze goes. You'd be very surprised if you were flying a kite and it suddenly flew into the wind!

Things are different in a plasma. In a plasma, we can have a neutral wind, an ion wind, and an electron wind *all going in different directions*.

This has led to what is known as the *wind shear theory* of sporadic E, in which, due to various magnetic and electric field influences, those winds all heading in different directions compress an area of E layer – usually it is the upper stretches of the E layer – into a dense layer of ionization.

There are things the wind shear theory predicts that turn out to be true, such as particularly frequent sporadic E over Southeast Asia, and particularly infrequent sporadic E over South Africa, both due to differences in the local magnetic field. There's a problem, though. We know that wind shear compression takes place over a period of about 100 seconds. So far, so good – that would explain why sporadic E seems to so suddenly materialize. However, at the densities involved, recombination would occur in approximately 10 seconds. In other words, the ionization would be evaporating almost as fast as it formed. Oops.

Enter the *wind shear plus metallic ion theory*. The E layer is composed mostly of oxygen and nitrogen plus some nitric oxide formed from the oxygen and nitrogen. That's the mix on which that 10 second number is based. However, instruments launched on rockets have found thin layers of iron and magnesium ions up there – probably leftovers of vaporized meteors. Metallic ions recombine much more slowly than gas ions. Could this be the key?

Maybe. However, this theory completely fails to explain a key characteristic of Es; Es is far more common in the summer than in the winter. There's no particular seasonal variation in that plasma wind shear, and no major seasonal variation in the metallic ion content. Of course, this means the theory also fails to explain why sporadic E dies off around the equinoxes. Finally, neither wind shear nor metallic ions decrease at night – and we know sporadic E is pretty rare at night.

It gets worse. If metallic ions are the key, then sporadic E should increase during or immediately after meteor showers. It doesn't.

It seems there must be at least one other mechanism, some seasonally variable mechanism, that occasionally either concentrates existing ionization or pumps extra ions into the E layer. There are some candidates at the moment, including dust particles, thunderstorms, and some as-yet-undiscovered characteristics or behaviors of the Earth's magnetic field.

There's an interesting NASA mission on the horizon that may shed some light on sporadic E. It's called the ICON mission, and is designed to study, among other things, the interaction of the troposphere and stratosphere with the ionosphere, and especially the connection of thunderstorms with the ionosphere. Thunderstorms do, after all, pump large quantities of ions skyward, so maybe there's something there.

How does one work a form of propagation that is usually only briefly present and is impossible to predict? Keep checking in on 6 meters! It's not for nothing that it is known as the Magic Band – it can be dead for weeks, then open up with fantastic propagation with no warning at all.

Sporadic E is primarily a 10 meter, 6 meter, and sometimes, 2-meter propagator, though at times it can work for lower bands as well. When 2 meters opens up, work fast – it's "call sign, locator, signal report" and on to the next, because those 2-meter openings usually become 2-meter closings in a matter of minutes. Lower frequency openings tend to last longer, and QSO's go at a pace more like regular HF contacts.

Sporadic E is most likely to occur around the solstices, especially the summer solstice, and prime sporadic E time in the northern hemisphere is May to August, with a peak in June. Time of day matters, too. When it occurs, sporadic E tends to peak around noon, then again around 7:00 pm. During those times, clouds of sporadic E that have diameters in the range of 500 miles are not uncommon.

The summer solstice is the longest day of the year – around June 20th – and the winter solstice is the shortest day of the year – around December 21st, so something about that long day seems to promote sporadic E. (We should note it's only the longest day in the Northern Hemisphere. The Sporadic E pattern reverses in the Southern Hemisphere.)

Amateur radio operators are, by the way, just about the only friends sporadic E has. Professionals see it as an annoying source of intermittent interference. That has played in our favor; a lot of research money has been spent on sporadic E, and not because we get all excited about 6-meter DX openings!

Ionospheric Scatter

"Scatter signals" happen when multiple levels of the ionosphere are reflecting the signal, or when pockets of different free electron densities form. The ionosphere is never "neatly organized" -- there's no exact altitude at which the F layer, for instance, begins. Think "shading" not a "border line." In scatter conditions, part of your signal might get bent a little bit by some of the ionosphere and a lot less by some a little higher, while another part gets bent a little more, then refracted back to Earth. The end result is scatter signals taking lots and lots of different paths into what would normally be the skip zone, so as the peaks and troughs of the waves reinforce then cancel each other out, the signal wavers.

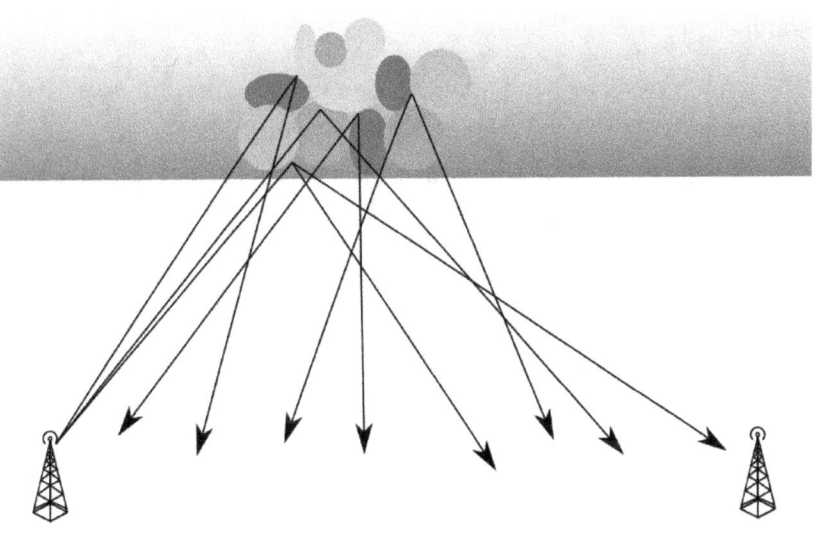

Ionospheric Scatter

It's precisely the same phenomenon we experience on VHF and UHF signals when we get multipath, but the end result sounds different because HF frequencies are lower and because we're typically using SSB amplitude modulation. Rather than the white noise bursts that are typical of VHF/UHF multipath, on HF we usually get rapid dips in signal strength.

Even though scatter leaves us with a weak, wavering, distorted signal, it does sometimes let us communicate over a distance too long for ground wave propagation but too short for normal sky-wave propagation. Sometimes it also lets us communicate on frequencies that are above the Maximum Usable Frequency.

Long Path/Short Path

There are plenty of times when a signal that can't make it to another point on the globe via the shortest path can make it by going "the long way around." It seems counter-intuitive, but when you consider that our signals regularly go much farther when they head out into space, a loop around the globe isn't so much, after all.

If you are using a non-directional antenna, there's not much you can do about long path/short path – your signal's going to get there however it gets there, or not. Let's say you have a rotatable beam, though, and you're trying to copy a signal coming from India. Your beam is most likely pointed roughly northwest, toward the short path to India. It's worth rotating the beam around to the southeast to try the long path.

(If you're scratching your head about those compass directions, get yourself a free downloadable azimuthal map centered on your QTH from https://ns6t.net/azimuth/azimuth.html.)

Chapter 8 - Getting Started in (Smart) HF DX

One great thing about ham radio is that you can approach any aspect of it as seriously or lightly as you want.

Some hams just like to sit in front of the radio once a week, twiddle the tuning knob, and see what comes out. If they like what they hear, they might strike up a conversation. If they happen to strike up that conversation with someone in Switzerland, well, great! If they never make a DX contact at all, that's great too! Station log? What station log?

If that's you, more power to you, and enjoy!

On the other end of the spectrum is the hard-core DX'er. These hams have all the gear, and they research the propagation conditions like a gambler studies the *Daily Racing Form*. They're linked into multiple online log systems. If there's a contest, they're in it. If there's a DX award, they've won it, or they're working on it and they are Not Fooling Around about it, either. If that's you, more power to you, too!

I think most of us are somewhere in the middle – we want to at least be good enough at this DX business to get regular contacts without hours of frustration listening to waves of static. We might be newly licensed Generals who are just dipping our toes in the HF waters, and a little puzzled about where to start.

As with most things in ham radio, successful DX'ing all starts with the gear in your shack – those radio waves aren't going to transmit themselves.

Transceiver

On the receiver side, modern top-of-the-line amateur transceivers are absolutely incredible machines. There's no imaginable way to manipulate, filter, and process an incoming signal that isn't available on these technological marvels. Things have changed by orders of magnitude from when I was a kid and sat in front of Dad's Hallicrafters short-wave radio in the garage. Things have also gotten dramatically more expensive.

HF transmitters, on the other hand, haven't really changed that much at all since the invention of SSB. They're still remarkably simple machines, and they do pretty much all that they've ever been able to do given the legal and physical limits of amateur radio signals.

To get a solid start in DX, though, you really don't need that $10,000 best-in-class miracle box. As I've mentioned elsewhere, hams were QSO'ing all over the world just fine long before the transistor was even invented.

I think the most important feature to look at when choosing a transceiver has nothing to do with its receiver sensitivity, or the amount of phase noise in the local oscillator. It's whether or not you like the radio and will feel comfortable operating it. After all, you're going to spend more than a few hours twiddling those knobs.

My transceiver is a 1980's vintage ICOM, and I like it. It has a big, bright LCD display with big letters. Most of the main controls are clearly marked knobs and switches, not items buried in a menu. (The miniaturized mic level and output power knobs seem to have been designed for a three-year old's fingers and are hidden under a ledge on the front of the radio, however, and that gets annoying at times.) The items that are on menus are fairly intuitive to get to and adjust, for the most part. It has a somewhat useful low resolution spectrum scope display – today's spectrum displays are far superior, but this one's workable. We also like that it covers 2 meters, which not all "HF" radios do; we can sit at our kitchen table and participate in our club's weekly net without switching radios.

Amplifier

Then there's the matter of power. Most HF transceivers these days are 100 watt output machines.

Is that enough?

Absolutely. People have earned their DXCC -- DX Century Club for worked 100 countries -- on QRP. That's 5 watts. I was just visiting a guy who has his WAS – Worked All States – his DXCC, and a wall full of contest certificates, and he's never in his life operated a box with more than 100 watts coming out of it.

Consider, too, the difference between 100 watts and the legal maximum of 1500 watts is only 12 dB, or 2 S units. Could those 12 dB make a difference sometimes? Sure! Do you need them? Nope. The longer I'm in the hobby, the more this truth becomes clear to me; it isn't the watts, it's the antenna and the skill and determination of the operator.

Speaking of antennas --

Antennas for DX

Ask any veteran DX'er about the key to DX success. They're almost always going to tell you, "persistence and a good antenna."

We've talked about a couple of limiting frequencies; the "critical frequency" and the "maximum useable frequency." I don't want you to get them confused. The critical frequency is reported by ionospheric observation stations. They blast a very high power signal that sweeps through a range of frequencies straight up, then listen for what comes back down. By timing the return, they know the height of the ionosphere layer that's returning their signal.

The critical frequency relates directly to the MUF, but it's not the same as the MUF. Unless you, too, are sending your signal straight up, the MUF for you is normally higher than the critical frequency, because your signal is going to hit the ionosphere at a different angle. The closer that angle is to the horizon where you're standing, the higher the MUF will be. That's why we really can't define an MUF for any given moment unless we specify "between which places."

Put simply, to create a skywave we want our signal to encounter lots of long-lasting free electrons up in the F layer. That means using a frequency high enough to bust through the absorbing E layer, or the D and E layers if it is daytime, but not so high that it also busts through the F layer. It would be very helpful if the F layer was thicker, wouldn't it?

The ionosphere also can only do so much when it comes to refracting our signal, so trying to have a signal make a 180° U-turn is asking a bit much!

In the next illustration, we see three different takeoff angles and the resulting skywave or lack of skywave.

Signal A is taking off straight up. It encounters nothing but the minimum vertical thickness of the ionosphere, and is on its way to a galaxy far, far away.

Signal B manages a moderately decent takeoff angle and encounters more ionosphere on its journey. It gets refracted some, but doesn't hang out in the free electron zone long enough to make it back to Earth. It's off to a different galaxy -- also far, far away.

Signal C takes off practically skimming the horizon. It runs into enough ionosphere to get bent back to Earth, and somebody's having a QSO with an Earthling.

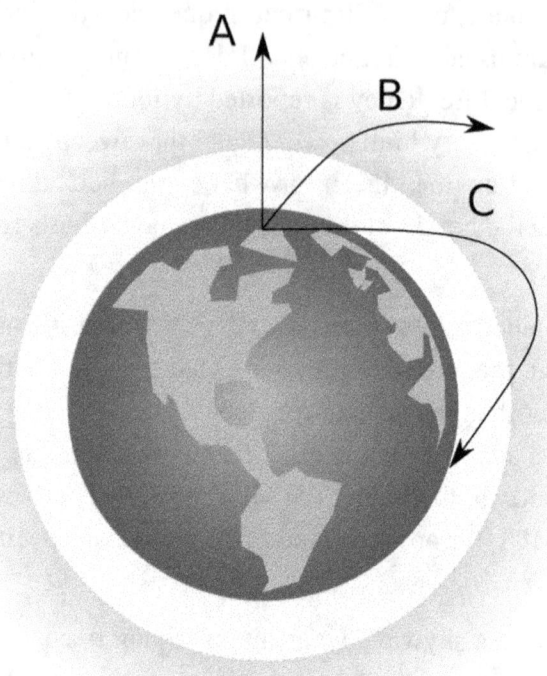

That's why the angle at which our signal leaves our antenna is so very important to successful DX'ing. It's really simple – as a general rule, most of the time, the lower your takeoff angle, the farther your signal will travel. "Lower" means, "closer to straight across the ground." So which antenna design gives us the most advantage for DX purposes?

That's one of the most hotly debated questions in all of ham radio! Before I get boiled alive in controversy, let me firmly state: we're talking here about "optimum" antennas, and this is not a complete list. It's also completely possible to achieve great things with less than optimal equipment. If none of what I describe is within your reach, don't let that stop you! Put up *something*, make it as good as you can make it, and get on the air.

While one of my favorite things I ever heard about antennas is, "When it comes to antennas, the short answer is 'there is no short answer,'" I'll try to be brief while I give you three broad categories to consider. Each has its pros and cons, and what's "best" for one person's situation might be the worst choice for someone else.

The main lobes of a *properly installed* ½ wave horizontal dipole will have a takeoff angle of around 35° above the horizon. The -3 dB beamwidth of the lobes will stretch from about 15° to 50° -- not bad at all. Even if you purchase a commercially available kit, the cost for a typical multi-band horizontal dipole is in the dozens of dollars.

Note, however, that often-challenging phrase, "properly installed." That means the dipole is sitting an honest ½ wavelength off the ground, with non-radiating supports at both ends, such as wooden poles or the ever-popular trees. That ½ wavelength off the ground is fairly easy to achieve if you have a couple of tall trees, but many of us don't have those. That leaves us pondering how to get a 40-meter dipole 66 feet up in the air, because anything short of that is going to push some portion of our signal straight up into the sky, with less of it going over the horizon where we want it. The antenna becomes an NVIS – Near Vertical Incidence Skywave – antenna. NVIS is sometimes great for crosstown and even regional contacts, depending on ionospheric conditions, but not great for DX. Do note: it *can* work, it just isn't optimal.

NVIS - Near Vertical Incidence Skywave

Horizontal dipoles don't radiate a whole lot off the ends, so they're somewhat bi-directional. If your antenna runs north and south, most of your signal will go east and west. Dipoles are not at all easy to rotate to a new direction – that's an afternoon's work even if you have places to hang it. Of course, they offer only a modest amount of gain vs. the fictional isotropic antenna.

The best thing you can do for your horizontal antenna for DX is to get it as high as you can – even higher than what the license exam told you, which was ½ wavelength.

Here are the optimum heights for antennas operating on various bands:

Band	Optimum Height
80 m	171 m/561 ft
40 m	68 m/223 ft
30 m	53 m/174 ft
20 m	38 m/125 ft
17 m	34 m/112 ft
15 m	29 m/95 ft
12m	24 m/79 ft
10 m	19 m/62 ft

Don't panic – those are *optimum* heights, not *necessary* heights.

Another choice is a ¼ wave vertical. The vertical pattern is, at least roughly, a cylinder centered on and parallel to the antenna. The vertical isn't pushing much signal up to the clouds. In terms of takeoff angle, we're still talking double digits. While the multi-band HF verticals available commercially are priced in the hundreds of dollars, in some circumstances they can end up being easier and more economical to set up than horizontal dipoles. Most models require radials to work properly, which will be a separate expense for you – then you get to figure out where and how you're going to bury or hide a hundred feet or more of wire in your back yard or on your roof.

Because properly installed verticals are not particularly directional in their horizontal patterns, and radiate much more signal horizontally than vertically, they're not optimal for NVIS work. Since their pattern is very much like a dipole's, only rotated 90°, they don't offer much gain, either.

That brings us to the heavy artillery of the DX world; tower mounted beam antennas with rotators. "Tower mounted" on a tall tower or a high hill, by the way – those optimum heights above still apply to horizontal beams. The advantages of these antennas are obvious – double-digit gain in both the vertical and horizontal planes. With a remote control rotator you can point them in the direction of your target without leaving the comfort of your shack.

The difference in takeoff angle between, say, 25° and 15° may not seem like much but it does offer significant advantage because as the takeoff angle gets closer to the horizon, the MUF goes higher. Here's the math.

Let's assume that all along the path between here and some point I want to reach, the critical frequency is 4 MHz. The formula for estimating MUF from critical frequency when the takeoff angle is worked in is:

$$MUF = \frac{f_{critical}}{\sin \alpha}$$

$f_{critical}$ is the critical frequency and α is the takeoff angle in degrees above the horizon. If we run this formula for a 25° takeoff angle,

$$MUF = \frac{4}{\sin 25} = 9.464 \; MHz$$

Well, the published MUF is 13 MHz, but my personal MUF is 9.464 MHz. What if I can get that takeoff angle down to 15°?

$$MUF = \frac{4}{\sin 15} = 15.454 \; MHz$$

Wow! Since most operators will look at the MUF and figure about 85% of that frequency is a good frequency with which to start, we've gone from our MUF being in the 40-meter band (30 meters is too high) to it being in the 20-meter band.

The disadvantages of big beams are just as obvious. Multi-band beams can cost thousands, and that's just the beginning of the expenses if you purchase and erect a large tower. Many homeowners associations and such fail to appreciate the beauty of an 85 foot steel tower topped by a giant Yagi antenna, as well.

Notice, by the way, that transmitter output power is nowhere to be found in that equation. If you were firing up a 100 kW transmitter, it would, but at the power limits of amateur operation, it does not. (That's not to say that there aren't other benefits to higher power, but a higher MUF is not one of them.)

Grounding

If you're about to set up your first HF station, please brush up on grounding techniques. They're important for your station's performance and for the safety of you, your loved ones and your property. There are three types of grounding in a radio station.

- Safety grounding. This is the grounding that keeps household AC and various DC voltages in your station equipment from going through you.

- RF grounding. This is what keeps the RF from your antenna from getting into all your equipment, creating distortion and possibly zapping your fingers when you touch your radio. Very importantly for DX, this is also what helps keep environmental noise out of your radio. You can't do much DX'ing if you can't hear any signals through the roar of static from all the newfangled lights and gadgets with which we've surrounded ourselves.

- Lightning grounding. Whether it's a simple dipole or a 200 foot tower, your outdoor antenna is a lightning rod. The lightning ground system keeps any lightning that hits the antenna and/or tower from coming into your house.

Depending on your antenna, there may or may not be a "ground radial" system as well – that's a separate issue, and does *not* provide lightning protection.

If you are the least bit shaky on this topic (you are) get the ARRL's book *Grounding and Bonding for the Radio Amateur*. It will be $23 well spent.

Strategy

I imagine most of us who have played with HF started with the same not-very-technical approach; turn on the radio, start tuning around, and see what happens. It turns out there are more productive approaches.

First, and this may be the most important thing I tell you about practical propagation, banish from your mind the possibility that "the bands are dead." No they're not. Propagation *somewhere* is *always* possible. Today, as I write this, conditions should be positively awful. The propagation reports and forecasts say every HF band higher than 40 meters is POOR right now, and 80 meters and 40 meters are only FAIR. The SFI – solar flux index – is 69, the sunspot number is 0 and has been for weeks. There's nothing in the space weather report that says, "Run to the radio!" Yet, according to dxmaps.com, QSO's are happening all over the globe on HF.

For *years* hams have been telling me "the bands are dead," yet – mysteriously – the ARRL keeps handing out those DX Century Club certificates. My startling conclusion: THE BANDS AREN'T DEAD. They're just not as lively as they used to be. One bit of good news? Less solar radiation does mean degraded propagation conditions – but it also means less recombination, and therefore, less signal absorption; we get stronger signals during solar minimums. We just get fewer of them.

So, yes, learn how to read the space weather, but remember the ionosphere is a complex and occasionally fickle beast – you won't really know what's possible until you turn on the radio and give it a try. As they say, you can't catch fish if you aren't fishing.

There you are – radio hooked up, antenna raised, grounding system installed, ready to chase some DX. Now what?

As a very general rule of thumb, daytime favors higher frequencies, nighttime favors lower ones, especially in the low points of the sunspot cycle.

To get more specific, consider where we are in the solar cycle, the season, and the time of day. Then consult the following charts. They show, by season and time of day, the most and least likely bands to be experiencing good DX propagation in peak, high, moderate, and low parts of the solar cycle. The darker the square, the better the propagation.

These are adapted from the book *The New Shortwave Propagation Handbook*, by Jacobs, Cohen and Rose, which is now, apparently, out of print.

Peak Solar Cycle – Summer -- Day

160	80	60	40	30	20	17	15	12	10

Peak Solar Cycle – Summer – Night

160	80	60	40	30	20	17	15	12	10

Peak Solar Cycle – Winter – Day

160	80	60	40	30	20	17	15	12	10

Peak Solar Cycle – Winter – Night

160	80	60	40	30	20	17	15	12	10

Peak Solar Cycle – Spring/Fall – Day

160	80	60	40	30	20	17	15	12	10

Peak Solar Cycle – Spring/Fall -- Night

160	80	60	40	30	20	17	15	12	10

High Solar Cycle – Summer -- Day									
160	80	60	40	30	20	17	15	12	10

High Solar Cycle – Summer – Night									
160	80	60	40	30	20	17	15	12	10

High Solar Cycle – Winter -- Day									
160	80	60	40	30	20	17	15	12	10

High Solar Cycle – Winter -- Night									
160	80	60	40	30	20	17	15	12	10

High Solar Cycle – Spring/Fall – Day									
160	80	60	40	30	20	17	15	12	10

High Solar Cycle – Spring/Fall -- Night									
160	80	60	40	30	20	17	15	12	10

Moderate Solar Cycle – Summer – Day									
160	80	60	40	30	20	17	15	12	10

Moderate Solar Cycle – Summer -- Night									
160	80	60	40	30	20	17	15	12	10

Moderate Solar Cycle – Winter -- Day									
160	80	60	40	30	20	17	15	12	10

Moderate Solar Cycle – Winter – Night									
160	80	60	40	30	20	17	15	12	10

Moderate Solar Cycle – Spring/Fall -- Day									
160	80	60	40	30	20	17	15	12	10

Moderate Solar Cycle – Spring/Fall – Night									
160	80	60	40	30	20	17	15	12	10

Low Solar Cycle – Summer – Day

160	80	60	40	30	20	17	15	12	10

Low Solar Cycle – Summer -- Night

160	80	60	40	30	20	17	15	12	10

Low Solar Cycle – Winter -- Day

160	80	60	40	30	20	17	15	12	10

Low Solar Cycle – Winter – Night

160	80	60	40	30	20	17	15	12	10

Low Solar Cycle – Spring/Fall -- Day

160	80	60	40	30	20	17	15	12	10

Low Solar Cycle – Spring/Fall – Night

160	80	60	40	30	20	17	15	12	10

Now take a look at your watch. What time is it? Just based on that, you can start to get at least a general idea of where you *might* start listening. If you have a particular DX target in mind, what time is it there? Ideally you and they are both in darkness.

Time	Sunrise	Daytime	Sunset	Night
	Gray line time!	National and regional contacts. International possible with good conditions.	Gray line time!	International contacts
Where to listen first	Usually 40 through 20 – 15 and 10 if they are open.	80 and 40 for local, regional and national. 20 and 17 for world-wide, sometimes 15, 12, and 10 in high sunspot times. In general, you're more likely to get overseas signals from the west early in the day, from the east later in the day.	Same as the sunrise gray line.	160, 80, 40 and 20 can all open up world-wide.
Notes	Listen before transmitting – don't step on the QRP folks!	If your rig covers 2 meters, remember transequatorial propagation can open up in late afternoon or evening.	Remember to work fast – you only have 45 minutes to an hour.	This is golden time for DX.

Let me emphasize again, all these are just starting points; at best, they are broad guidelines. You won't really know what bands are working for *you* *right now* until you turn on your radio.

Beacon Stations

Next, you might give a quick twist of the dial to some beacon stations. Any properly licensed amateur may operate a beacon station – but then, those properly licensed amateurs may also *stop* operating their beacon stations. The most dependable beacons tend to be those that are part of the NCDXF/IARU Beacon Project. NCDXF is the Northern California DX Foundation, and IARU is the International Amateur Radio Union.

At present there are 18 beacon stations in the Beacon Project located all around the world. All the stations use the same set of frequencies: 14.100 MHz, 18.110 MHz, 21.150 MHz, 24.930 MHz, and 28.200 MHz. The transmissions of the stations are coordinated, so as one station stops transmitting on, say 14.100, another station starts transmitting on 14.100 while the first station switches to 18.110 MHz. Each transmission takes 10 seconds and consists of the station's call sign, sent in Morse Code, followed by four one-second dashes. The call sign and the first dash are sent at full power, 100 watts. The second is at 10 watts, the third at 1 watt, and the last at 100 milliwatts. This gives you at least a rough gauge of the quality of the signal path and how much power you might need to use to reach the beacon's region.

You haven't learned Morse Code yet? Not a problem; the NCDXF's web site has a constantly updated chart of which station is transmitting on which frequency: http://www.ncdxf.org/beacon/index.html.

Checking the beacons will give you at least some sense of which bands are propagating and which are not. If you have a directional antenna, you'll also know where to point it.

Another relatively new and rapidly evolving tool for DX is the *reverse beacon network*. Rather than sending out signals, the "reverse beacon" stations listen for your signal, then report, via the internet, that they have heard you. The "reverse beacon" stations are known as *skimmers*. They're broadband software defined radios. Their output is fed to a computer running skimmer software that "listens" to the output, decodes it, then posts what it has heard on http://reversebeacon.net. You can look yourself up on the web site and can get at least some sense of how your signal is propagating. ("Some" sense because the skimmers are, of course, volunteer operations, and there's no grand plan for their geographical distribution around the globe.) At present the skimmers can decode CW, RTTY, and PSK31 modes, with more in the works. For a demonstration, go see EI2KC's video at:

https://www.youtube.com/watch?v=xhAqZV0RH4E.

You can also reverse beacon yourself. Go to:

http://websdr.org

Pick an SDR – Software Defined Radio -- station near you on that site. Tune the virtual radio to a clear frequency, set your transceiver to that frequency, and once you've assured that the frequency is not in use, say something like "AF7KB testing." Did you see your signal show up on the SDR display? You're getting out! Pick another SDR farther away and repeat as necessary. (If you're not familiar with websdr, it's a site that aggregates the outputs of a bunch of SDR's all over the world. Very easy to learn to use – you'll have it down in minutes.)

Finally, you have some choices to make about what mode to use. There's no avoiding the fact that until the solar cycle starts picking up again, HF propagation is not going to be magnificent much of the time. Consider something besides SSB. CW is still a tremendously powerful tool, as are the WSJT modes, such as FT8 or its more chatty cousin JS8CALL.

Propagation Forecasts

Without a propagation forecast in hand, that's about as exact a plan as one can have; however, it's not at all hard to get the latest space weather report and even to find excellent propagation forecasts.

It starts with getting a good sense of general conditions. Fire up the web browser and navigate to http://www.hamqsl.com/ for the latest space weather reports and general propagation forecasts.

If you were going to look at only one value, you'd be best served by either the sunspot number or the solar flux index – but remember, even with dreadful numbers, propagation is possible! (It just may take some patience.)

For a more detailed look, there are ham operated web sites that will give you current data on what bands are open where.

Here are some sites to check out:

https://www.dxmaps.com/spots/mapg.php
http://hflink.com/propagation/

There's also an incredibly cool, if, perhaps slightly (?) nerdy, app from NASA ("Space Weather") that gives you nearly real-time pictures of what the Sun's up to. Not terribly useful for ham radio, but, wow, those pictures are amazing!

DXLab Suite

You can also get software to do your own propagation calculations. Many programs like that have come and, sadly, gone. Web searching for them will get you lots of messages that go something like, "I am no longer supporting this software. Use it at your own risk. It needs to run on Windows 95 or earlier." (I might be exaggerating – but not much.)

However, there is one great one that is kept constantly up to date, is fairly easy to learn, is extraordinarily useful, and is free! That's the DXLab Suite from David Bernstein, AA6YQ. The program is for Windows only. Download at

http://www.dxlabsuite.com/dxlabwiki/InstallLauncher

DXLab Suite has many useful capabilities. It's a suite of programs whose uses range from propagation prediction to operating your computer-controllable transceiver. We won't even begin to cover all the ways of using DXLab Suite here.

Because it has lots of functions, there's certainly a bit of a learning curve to using DXLab Suite, but there's an active online user community and tutorials. Here's one way to use it.

Let's say I want to QSO with the fellow in Ireland I mentioned earlier, EI2KC.

I open the parts of DXLab Suite called PropView (Propagation View) and DXView. Here's the window of DXView:

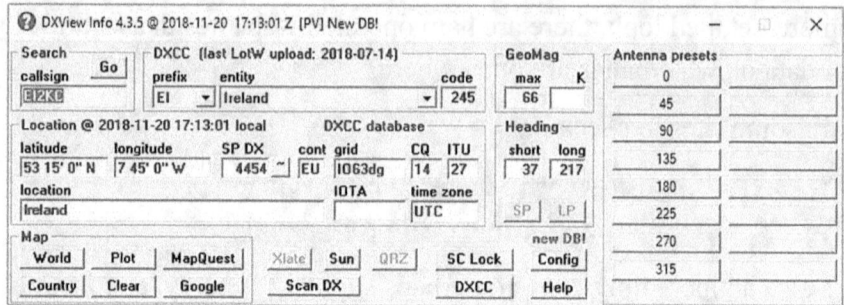

The only entry I made is "EI2KC" in the call sign. The program filled in everything else, including his latitude, longitude, grid square, and even the headings for the short path and long path to him. It did it almost instantaneously, too.

Don't have a particular call sign in mind? Not a problem! In the lower left corner you see the "Map" section of DXView. Click on "World" and a map like this opens:

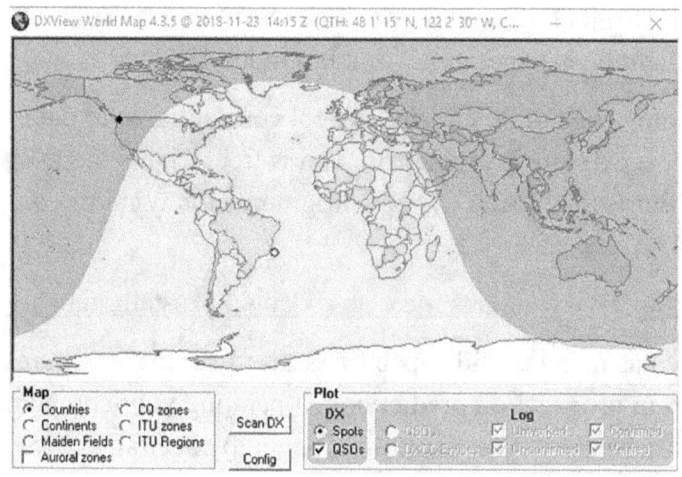

Click on a destination, and those coordinates auto-fill in the DXView window.

The coordinates in DXView transfer automatically to PropView, which is the part of the program that's going to create your propagation forecast.

The PropView window looks up the latest space weather numbers with the click of a button and puts them in the "Conditions" section: the Solar Flux Index, the Sunspot Number (in this case the *Smoothed Sunspot Number*), and the K index. You pick the "Avail" number; that's a measure of, basically, how picky are you about the probability of the path being measured actually working. You also pick the desired SNR, the Signal to Noise Ratio, and/or the mode you want to use. SSB demands less noise than CW, for instance.

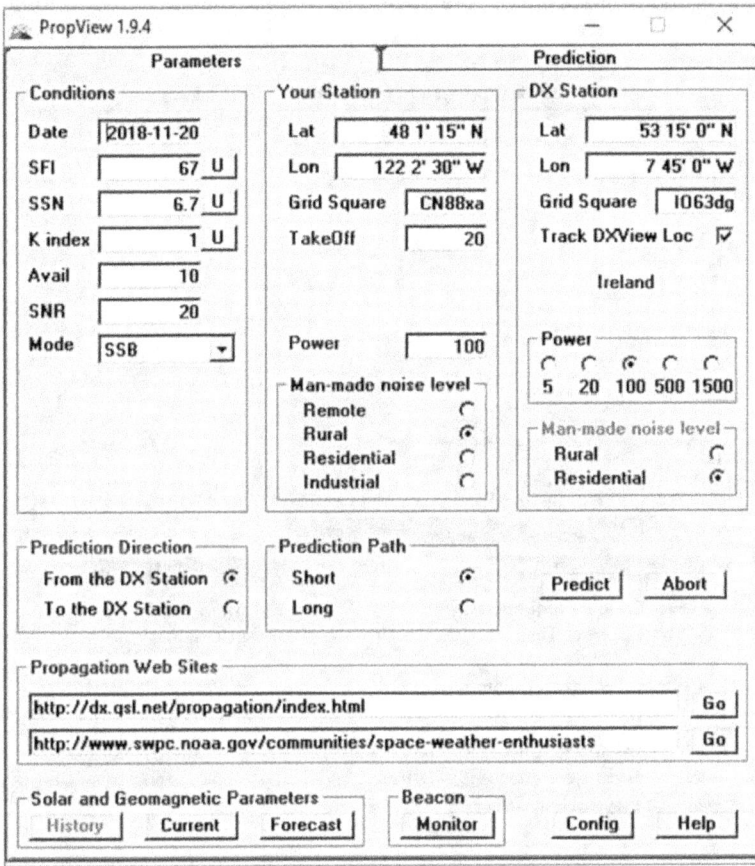

There are also choices you can make about the direction of the prediction – from you to them or vice versa; the noise level at your QTH, the noise level at the target QTH; your power level and theirs; and, long path or short path.

Then you click the "Predict" button and the program generates a propagation prediction for the path between you and the target you have chosen. It looks like this:

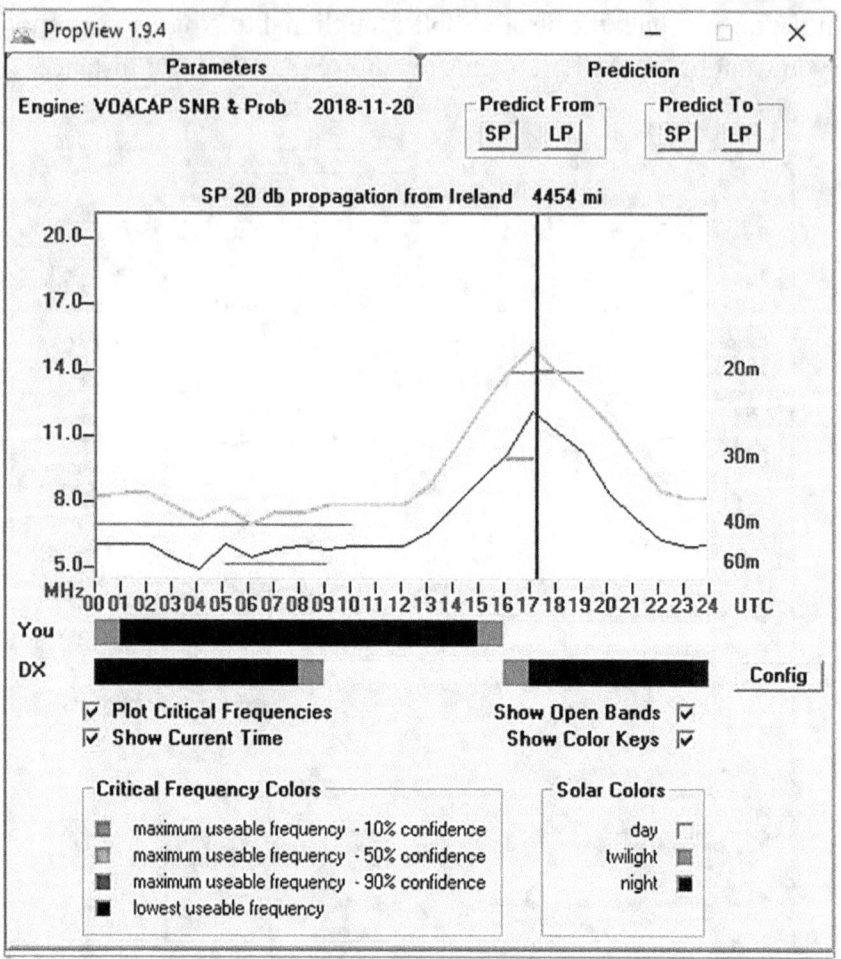

The graph covers 24 hours. The X axis is time-of-day in UTC, with the vertical line showing the current time. The bars beneath the time scale indicate day and night at your QTH and at the target.

The vertical axis is frequency. The left hand vertical axis is calibrated in MHz, the right in wavelength.

The line graphs indicate the predicted critical frequency – not to be confused with the MUF. The top line predicts the critical frequency with 50% confidence, the bottom line predicts it with 90% confidence. The program also generates a critical frequency prediction with 10% confidence, but because I have "Show Open Bands" checked, it isn't displaying that one, since it's up in some closed band.

Sometimes the chart will show you the LUF as well, but in this case it was below the ham bands.

The horizontal lines on the chart are the ones you want to see a lot of. They indicate band openings, and they get wider as your chances on the band get better. You can see that at the moment I made this prediction, I had about two hours of 20-meter opening left before the band was predicted to close.

(It might be useful for you to know that the actual graphs produced by the program are color-keyed – if you're holding the print edition of this book, you're obviously not seeing all the colors, which make these predictions far easier to interpret.)

DX Toolbox

Another propagation tool called DX Toolbox takes a slightly different approach. DX Toolbox is available for Windows or Mac. Like DXLab Suite it incorporates a vast array of functions, including a real-time satellite tracker. However, for our purposes, I'll focus on the Propagation Window.

To set up DX Toolbox, you enter your latitude, longitude, and altitude in the Preferences window. From there, you can open several different displays of current space weather conditions and propagation predictions. You can, for instance, see a 24 hour propagation display for a particular band from your QTH to a chosen country, or a 24 hour estimation of the MUF/LUF on that path.

Opening the Propagation Window shows you a map of the world, usually with some areas blacked out to indicate "reception is not possible" to and from that point on the map. It already knows the solar flux index. You fill in your desired frequency and your station's power.

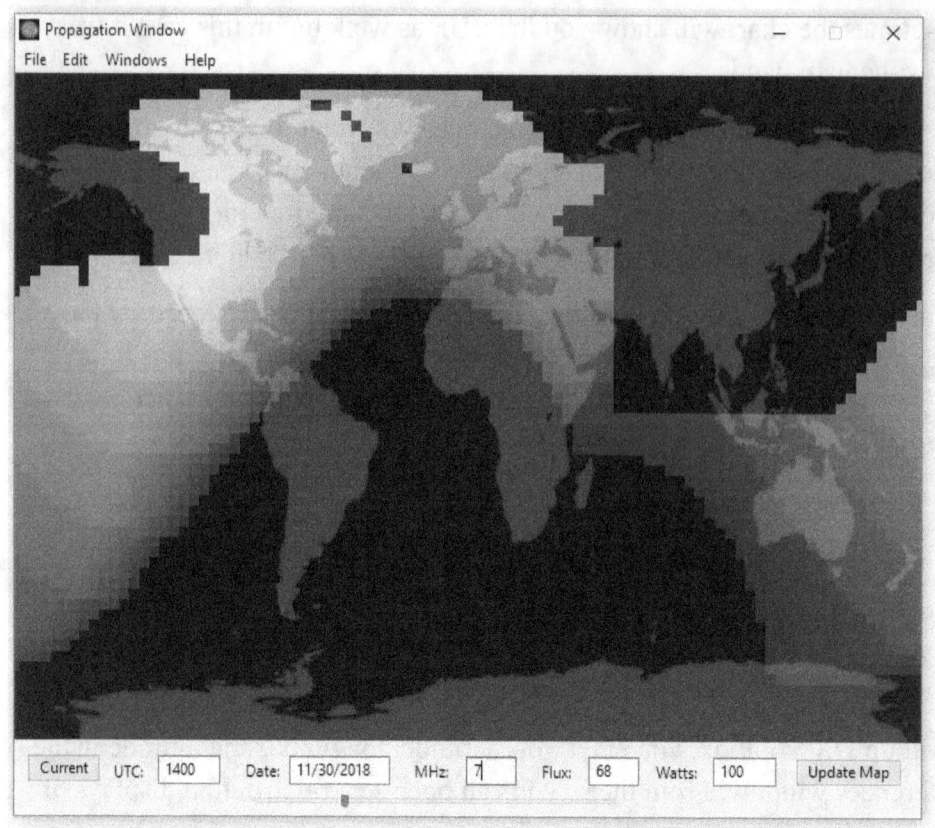

DX Toolbox is shareware. Present price is $24.99 to register the software. It's available from:

https://www.blackcatsystems.com/software/ham-shortwave-radio-propagation-software.html

PropLab Pro

The heavyweight propagation modeling program is PropLab Pro.

Most of the other propagation prediction programs are using the algorithms from VOACAP, a propagation program originally developed by the Voice of America. Put simply, VOACAP looks at a handful of space weather indices and creates a general prediction of propagation on a particular path. It's really quite a useful program IF you understand its limitations. Think of it this way; it's not telling you "These are the conditions right now," it's saying, "Given what I know, and the path you've chosen, this is how conditions most likely will be for the next little while – or so."

PropLab Pro operates differently. For one thing, its model of the ionosphere is more fine grained than what VOACAP based programs generate. For another, PropLab Pro is "ray tracing" software. It actually builds a model of how your signal is predicted to propagate – it even builds it in 3D.

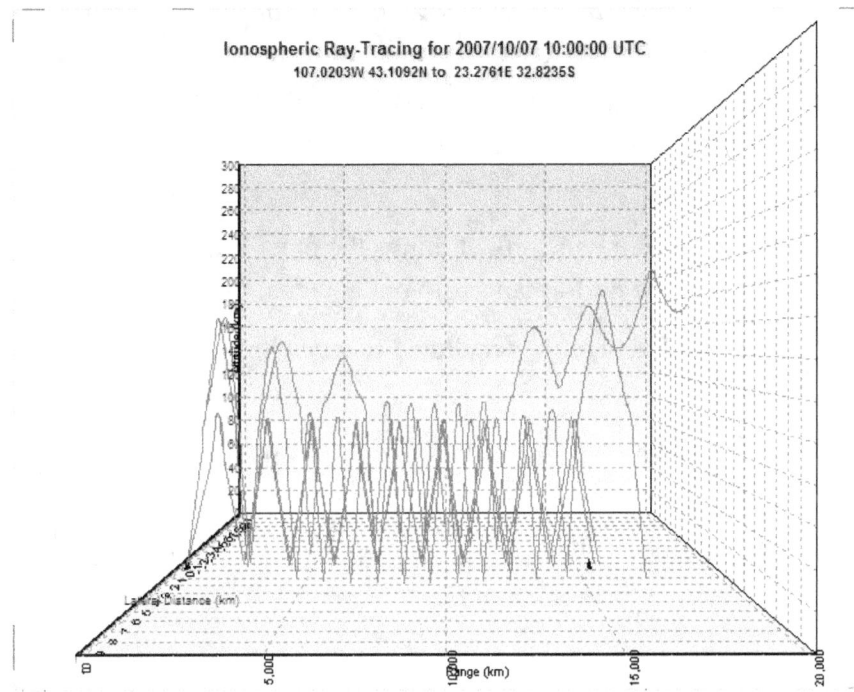

Ionospheric Ray-Tracing for 2007/10/07 10:00:00 UTC
107.0203W 43.1092N to 23.2761E 32.8235S

That model is based on the "2007 International Reference Ionosphere." I'll quote a brief bit of PropLab Pro's user manual – which has to be the most breezily conversational and genuinely useful user's manual of any software I've ever seen – for an explanation:

As you may have heard, Proplab-Pro comes equipped with the 2007 International Reference Ionosphere (also known as the IRI). There have been many different versions of the IRI throughout the years. Every so often, some of the worlds smartest scientists and other intelligent beings get together to help describe the nature of the Earth's ionosphere in mathematical terms. All of the ideas, theories and mathematical innards that form the basis of our understanding of the Earth's ionosphere is then compiled together by the working group whose task it is to create the next International Reference Ionosphere. Their work is then passed onto another small group of equally smart people who know no other language than that spoken by computers. Their job is to make a working set of computer codes that can be used to fully describe the Earth's ionosphere in the best possible terms, using one of the oldest and most archaic computer languages known to man. This is akin to writing a modern masterpiece entirely in Latin. Their completed project is then tactfully named the International Reference Ionosphere and is pre-fixed with the year that their work was completed. At the time of this writing, the latest and greatest version of the IRI was completed in 2007. We have kindly retranslated and integrated that code into Proplab, with only one of our programmers suffering premature aging in the process.

PropLab Pro takes into account all sorts of things the other programs generally do not, such as the performance of your antenna.

The program is capable of generating an almost overwhelming array of analyses.

As you'd expect, with all those superpowers, it's more expensive than other programs. The current price is $240.00.

http://shop.spacew.com/index.php/product/proplab-pro-hf-radio-propagation-laboratory/

Chapter 9 – Meteor Propagation

Most folks seem to think of meteors as somewhat rare phenomena, but there's a constant rain of meteors into our atmosphere. Estimates vary, depending on the methods used to make the estimate, but around 50,000 tons of meteors hit our atmosphere in a year; that's a couple of hundred pounds of space rocks incoming every *minute*.

We don't ever see the vast majority of them. They're dust particles. The "shooting stars" you see on a dark, clear night are mostly pea sized chunks. (Baby peas, at that.) Still, between 20,000 and 80,000 meteors per year are larger than 10 grams.

Meteors enter the upper atmosphere at speeds ranging from 5 to 50 miles per *second*. That's a lot of kinetic energy. Military tanks firing kinetic energy penetrator weapons manage a projectile velocity of 5,600 feet per second. The slowest meteors top 16,000 feet per second. When they hit the atmosphere they give up that kinetic energy in the form of heat. Each of those meteors, regardless of size, leaves behind a trail of ionization as that heat ionizes the atoms in the atmosphere.

Most meteors would barely qualify to be called dust, and don't leave a useful number of free electrons behind, but there are still 1,000,000,000,000 (10^{12}) a year that do leave enough ions to refract a radio signal. Because of their tremendous speed, a meteor that weighs in at a mere 0.10 microgram is enough to generate a useful trail of ions.

The heaviest concentration of ions usually ends up at the level of the E layer; as the meteor descends, it encounters thicker and thicker air, and the E layer is where the meteors usually vaporize, never reaching the D layer level.

The heat of the meteor's vaporization ionizes air around it, and the meteor also leaves behind a trail of vaporized metal – more ions. All that adds up to dense pockets of free electrons, so 10 meter, 6 meter, and 2-meter signals -- frequencies that would normally shoot straight through the E layer -- are, instead, refracted or scattered back to Earth.

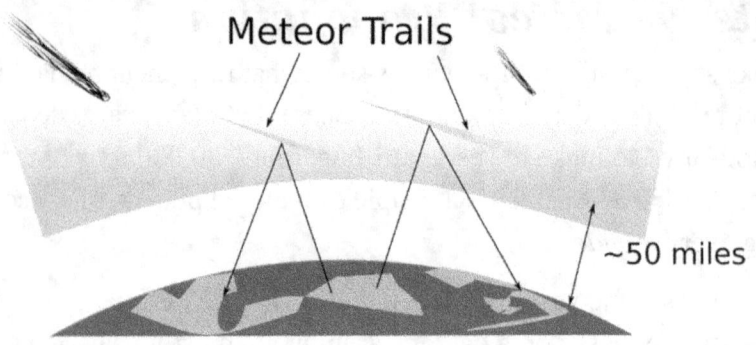

The density of the air in the E layer that vaporizes the meteorite also contributes to the rapid recombination of the ions and free electrons that are formed, so any given occurrence of meteor scatter is a brief event, indeed. Think in terms of seconds or, maybe, on rare occasions, a minute of communication.

How the Pros Use Meteor Scatter

I have to confess that when I first learned about meteor scatter propagation, I thought it sounded fun but ultimately rather pointless – one of those "wouldn't it be cool if this worked" ham radio activities with no practical application. Was I ever wrong! Hams have had some remarkable communications achievements with meteor scatter – a few have even achieved the WAS (Worked All States) award using only VHF. That took a lot of meteor and tropospheric scatter as well as sporadic E skip. (It also helps to live in the Central US for this feat.)

Successful use of meteor scatter isn't limited to the ham world, either. It turns out the Canadian Defence Research Board deployed a very effective communication system known as JANET between remote Prince Albert, Saskatchewan and Toronto, Ontario back in the early 1950's. That site in Prince Albert was one of their key radar research stations and there were no phone lines between there and Toronto.

According to Peter A. Forsyth, who headed the project, it was named JANET after the double-faced Roman god Janus, who looked both ways. His very enjoyable account is here: http://www.friendsofcrc.ca/Articles/Forsyth-Janet/Janet-PeterForsyth.html

Later, in the 1960's, NATO's Supreme Allied Headquarters Europe built another system that facilitated strategic communication among the various

European nations. The US Defense Department continues to develop and experiment with meteor scatter systems, as well. The US Department of Agriculture's entire SNOTEL system (Snow Telemetry) of remote snow pack monitors sends all its data via meteor scatter. Why all this interest in an exotic propagation mode in an age when we have instant, reliable communications via satellite?

First, satellite communication isn't so dependable in those high latitudes – remember all that aurora activity around the poles? Second, satellites are expensive, while ground based systems are, relatively speaking, bargains. Ground based systems are repairable. If a radio breaks on a satellite, that's that. If a radio breaks on the ground, you send out a tech in a van. (In the case of SNOTEL, it's a tech on a snowmobile.) The military has a particularly keen interest in the "small footprint" of meteor bounce signals.

We can take some cues from how those "grown up" systems worked.

In the Canadian system, for instance, the radar station's unit listened for a beacon signal from Toronto on 90 MHz. As soon as it heard that beacon, it triggered a tape that sent a quick burst of data on RTTY. (At up to 1300 words per minute!) Then the radar station unit waited for an acknowledgement signal from Toronto. If it got the signal, it could send the next burst of data. No signal? Rewind the tape and be ready for the next chance.

Many details of the other systems are still classified, but so far as we know they all worked – or, maybe, still work – on much the same basic system.

Even though it sounds like a contradiction in terms, we can say that meteor scatter is *dependably intermittent.* We don't know when it will occur, nor for how long, but it will definitely occur, so we can build our strategy around that.

Working Meteor Scatter

Let's look at the elements of success. First, there's pre-arrangement. Both parties are "in the know" about when and on what frequency the communication will occur. Both parties know where the other is located, so they know where to point their antennas. Finally, there's speed of transmission and repetition of those transmissions.

For us, the equivalent of the Canadian system's beacon station is that other ham out there sending "CQ scatter, CQ scatter, KG7NVJ KG7NVJ break!" When we hear that, we need to jump on it quickly and briefly. "KG7NVJ AF7KB AF7KB break!" Chances are good, that's all the time we'll have. Experienced operators often have a portable audio recorder with them so they can go back later and catch the call signs that went flying by at Mach 3. (That's also a popular strategy for satellite operators.)

Some folks still use SSB phone – not FM -- for meteor scatter, but the weak signal digital modes are more popular and can be very effective; especially FSK441, which was designed specifically for meteor scatter. One of the most dependably effective modes is, not surprisingly, CW.

As always, more power makes for more dependable communications, but on 6 and 10 meters you can get a lot of contacts with 100 watts and a portable dipole or Yagi. For 2 meters, most successful operators use high-gain beam antennas and high power amplifiers.

If you're new to meteor scatter, you'll want to try 10 meters and 6 meters first – 2-meter propagation disappears *fast*, while 10 and 6 tend to hang in a bit longer. Twist that knob to 6 or 10 meters and listen for "CQ scatter!"

Better, hook up your computer to your radio and get the WSJT suite of "weak signal digital" protocols. The MSK144 protocol is specifically designed for meteor scatter – or as the pros know it, *Meteor Burst Communication*, because communication depends on sending quick bursts of information in the brief moments of an opening.

Here's the geometry of the situation:

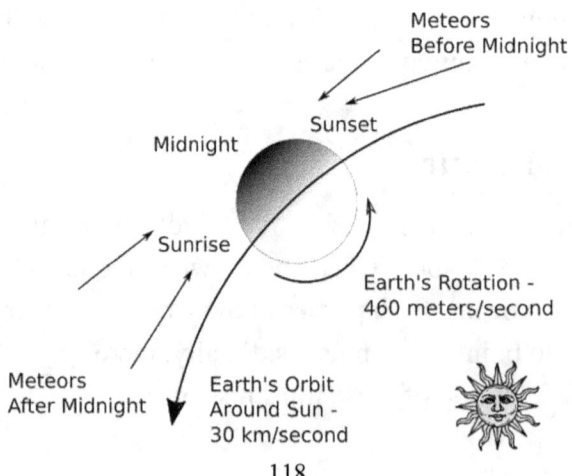

If possible, point your beam "forward" relative to the direction the Earth is traveling. That's only possible after midnight.

There are many more meteors between midnight and sunrise than any other time, because the Earth is busily plowing into them at a brisk 30 km/second. Before midnight, the meteors have to chase Earth down. Given that meteor chasers tend to be playing between pre-dawn and about 0900 on Saturdays and Sundays, those are your prime times.

The average number of meteors varies predictably through the seasons, too. Seasonal variations peak in August, with the minimum in February; our orbit takes us through a region of denser material in August.

Of course, the very best time for meteor scatter is during a meteor shower, when the count of visible meteors per hour can climb well past 100. On rare occasions, that number can top 1,000. Meteor showers occur when the Earth passes through clouds of cosmic debris that are hanging out in Earth's orbit – remnants of those cosmic litterbugs known as comets. Times, dates, and even intensity of meteor showers are all quite predictable, and you can look them up on sites such as:

http://space.com

During a meteor shower, you'll point your beam at the constellation out of which the meteors seem to be coming, known as *the radiant*. In August, that's Perseus. (The app called Heavens Above, or similar apps, will help you find that in the sky.)

Chapter 10 – Auroral Propagation

If you place a magnet under some paper and sprinkle on some iron filings, you'll see the magnetic lines of force surrounding the magnet.

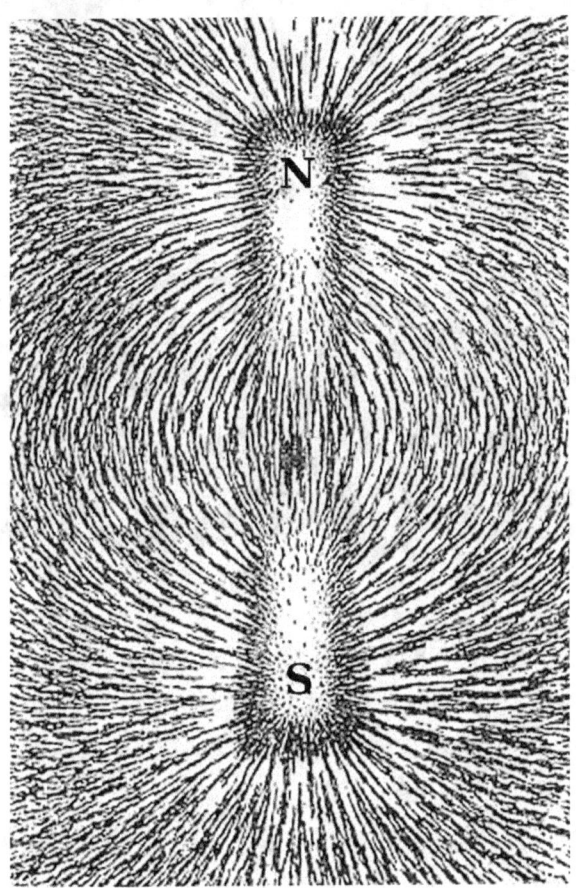

Since the Earth is a big magnet, it has lines of magnetic force surrounding it too.

Remember, magnetic lines of force can be deformed by other magnetic lines of force. Intense blasts of the magnetic force of the solar wind make those magnetic lines of force around the Earth move around in space. You'll recall the picture of the magnetosphere from an earlier chapter:

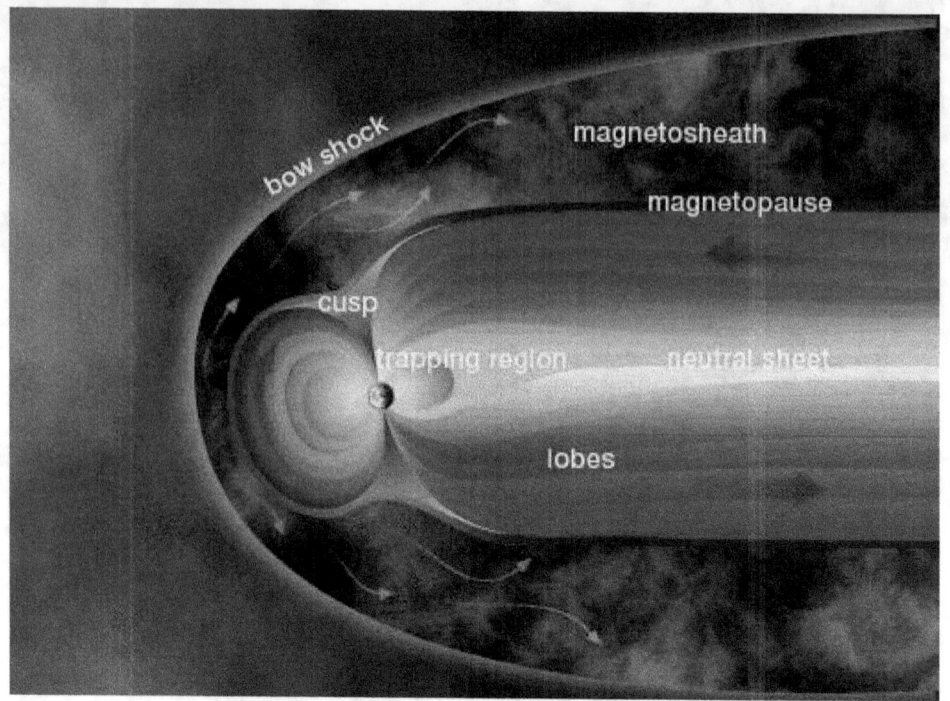

Earth's Magnetosphere

In quiet sun conditions, very few of the particles that make up the solar wind penetrate the ionosphere. What few get in tend to get in at the poles – you can see why by looking at the shape of the magnetic field above. They mostly stream around the Earth and go on their merry way, much the way water in a river would flow around a rock.

Here's a very necessarily simplified view of what happens under not-so-quiet sun conditions, such as when a solar flare aimed our direction occurs. The solar wind speeds up and the volume of ionized particles it is carrying increases. With the increase in strength of the solar wind, the magnetosphere changes shape. The result is that the area labelled above as the *cusp* opens up. That allows the solar wind to enter the upper atmosphere, bringing with it that stream of ions. Obviously, those ions and free electrons are streaming in at the Earth's poles. However, that stream of ions and free electrons alone, even though it is traveling at solar wind speeds of 100's of km per second, still would not provide adequate energy to produce auroras.

Back in 1999, the FAST (Fast Auroral Snapshot) NASA spacecraft provided data that showed that it is the electric fields aligned along those magnetic lines of force that accelerate those already fast-moving particles earthward.

It's that acceleration that pushes the energy levels over the top and creates the auroras.

As you would expect, all that electrical activity has a profound effect on propagation. In times of intense aurora activity, HF propagation over the poles is wiped out, and is only possible in limited latitudes – if at all.

VHF, on the other hand, can be scattered off all those free electrons in the aurora zone creating long-distance 6-meter and 2-meter hops. Propagation distances of 500 miles are not uncommon, and the current official record is 2,178 km (1,353 miles.)

Auroras are an almost constant phenomenon above about 67° north magnetic latitude, though sometimes they're so weak as to be pretty much invisible. 67° is near the top edge of Canada – not easy territory to access. As geomagnetic disturbance increases, which is indicated by an increase in the K index, the auroras get brighter and creep south. At a K index of 9, they can often be seen as far south as 48° north magnetic latitude. Extraordinary conditions have very occasionally created auroras visible from northern Texas.

All those processes and values are mirrored in the southern hemisphere.

Note that key phrase, "magnetic latitude." You'll recall, that doesn't necessarily match up with "geographic latitude" because the magnetic poles aren't in the same place as the geographic poles.

Find the Auroras

Here's a table of some major US cities with their geographic latitude and magnetic latitude.

City	Geographic Latitude	Magnetic Latitude
Atlanta	33.7	44.5
Boston	42.3	51.7
Chicago	41.8	52.2
Dallas	32.8	42.7
Denver	39.7	48.3
Great Falls, MT	47.5	54.9
Los Angeles	34.0	39.8
Minneapolis	44.9	55.1
New York	40.7	50.6
San Francisco	37.7	42.5
Seattle	47.6	52.7
St. Louis	38.5	49.2
Washington, DC	38.9	49.1

Here's how far south the auroras go, generally, as the planetary K index (K_p) climbs.

K_p index	Magnetic Latitude
$K_p = 0$	66.5
$K_p = 1$	64.5
$K_p = 2$	62.4
$K_p = 3$	60.4
$K_p = 4$	58.3
$K_p = 5$	56.3
$K_p = 6$	54.2
$K_p = 7$	52.2
$K_p = 8$	50.1
$K_p = 9$	48.1

The northern edge of the lower 48 US states is roughly magnetic latitude 53° north, so it takes a K_p (Planetary K) of at least 6 or 7 for us to see auroras down here. Our brothers and sisters in Alaska will see auroras with a K_p of a mere 3 or less.

Auroras generally occur between 50 and 80 miles up, but can extend as high as 250 miles above the surface. That altitude is important; we don't have to be sitting directly under the aurora to take advantage of auroral propagation. At least in theory, at 50 miles high, the aurora can be about 500 miles north of us and still be workable. At 80 miles, the distance increases to about 800 miles, and if the aurora extends up to 200 miles, that distance could be 1000 miles!

For aurora predictions, navigate to the Space Weather Enthusiasts page at the National Oceanic and Atmospheric Administration (NOAA):

https://www.swpc.noaa.gov/communities/space-weather-enthusiasts

There, you'll find all the space weather numbers and an Aurora Forecast Map.

The forecast map shows you the probability of *visible* aurora with color coding. On the forecast shown, it isn't an exciting aurora day – all the shaded area around the north pole is green, indicating a 20% or lower chance of visible aurora in those high latitudes. If it was red, it would be 90%.

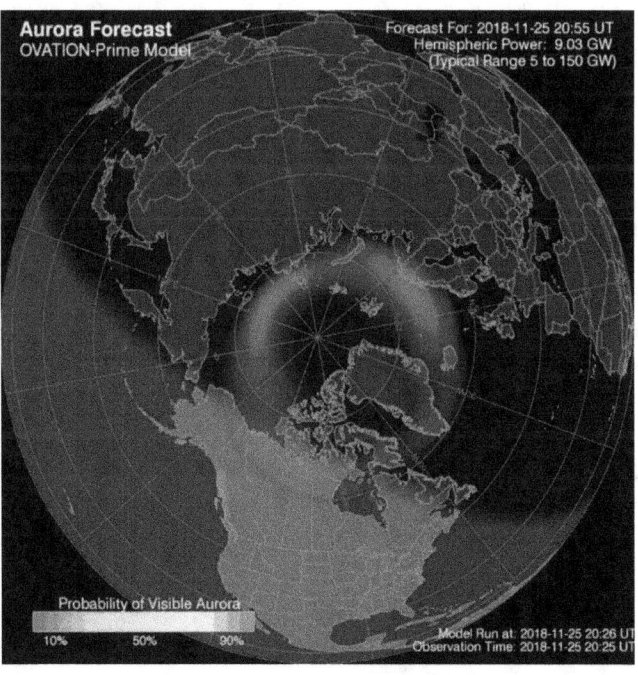

Working Aurora Skip

Equipment for working aurora propagation can consist of a 100 watt transceiver capable of working 6 meters and/or 2 meters, and a high-gain beam, such as an 8 or 9 element Yagi. Point that beam right at the aurora.

Statistically, your best choice for an operating band is 2 meters, though 6 meters can work, as well. Very occasionally, 1.25 meters will open up and even 70 cm.

Preferred modes are CW and SSB. Most aurora signals are very distorted – SSB phone often sounds like a fluttery, raspy, whispering ghost has taken up residence in your radio when it is intelligible at all, and even CW loses its tone and becomes more like static bursts that have been put through a blender. Digital modes do not work at all well for auroral propagation.

All that distortion results, in large part, from the fact your signal is passing through very strong magnetic lines of force that twist and tear your nice neat lumps of RF in ragged random shapes.

There's more. Recall, too, that aurora are caused by particles – electrons and protons – being accelerated at enormous velocities down toward the poles. We're reflecting our signals off an oncoming wall of particles, and that means that in addition to all the distortion, our signal (and the signal we're trying to hear) will usually be Doppler shifted up in frequency, by a matter of 500 to 1000 Hz or so. Those values can go higher during major geomagnetic disturbances, and they can also go negative. Practically speaking, that means a lot of VFO jockeying unless you program in an offset – a practice which may or may not pay off.

Operating procedures are essentially the same as regular DX; "CQ CQ CQ AF7KB Alpha Foxtrot 7 Kilo Bravo CQ CQ CQ," then listen for the reply. Signal reports often get an "A" for aurora appended to them; so, instead of "You are 5 7," it's "You are 5 7 A."

Keep an eye on that K index and when it hits a promising number, tune up the rig to 2 meters, point your beam North, and give auroral propagation a go.

Chapter 11 – Space Propagation

As amateurs, we use space propagation – propagation through outer space – for Earth-Moon-Earth communications and, to a certain extent, when we use amateur satellites.

If our space signals traveled only through outer space, this would be a very short chapter! For practical purposes, propagation through outer space is that form of propagation that turns out to be so very rare down here on Earth; plain old line of sight.

However, all our space propagation begins and ends here on Earth, and that means a trip through the troposphere, the stratosphere, the ionosphere, and possibly even the exosphere, then a return trip back through all that.

Even though with space propagation, we're just firing straight through the ionosphere, purposefully using frequencies that won't generate skywaves, here's what can happen to our signal along the way.

- Attenuation due to absorption. The cures here should be apparent: more watts, higher frequency, and a higher gain antenna.

- Refraction, causing a change in the angle of signal arrival. This is not very significant for us. Our satellites are in low earth orbit, which is only a few hundred miles above the ionosphere. You're probably not running a 20-meter radiotelescope style dish antenna with a super-tight beam. Your Yagi's beam isn't so narrow that a little bending will make it miss a satellite – and the Moon's a pretty big target.

- Scintillations. Rapid variations of the signal's strength and phase due to irregularities in the ionosphere. There's not a whole lot we can do about this except try another frequency. The higher the frequency, the less susceptible to scintillations it is, so you know the direction to look!

- Propagation of the signal at speeds slower than the "normal" speed of light. This isn't much of a concern for us, but is a big concern for the GPS network engineers. If your GPS ever showed that you were driving through the middle of a lake when, in fact, you were on a freeway, the ionosphere was acting up and delaying a signal from one or more of the GPS satellites.

- Faraday rotation. Magnetic fields can alter the polarization of electromagnetic waves, and that's called Faraday rotation. Our signals must pass through the Earth's magnetic field, and the angle through which they intersect that field determines, in part, the polarization that will come out the other side. As a very general rule of thumb, East-West propagation will tend to remain linearly polarized, North-South will tend to become circularly polarized. Propagation diagonally across the field will tend to be elliptically polarized. This can be the cause of those wonderful (?) slow fades in and out as the received signal's polarization aligns with our linearly polarized HF antenna, then goes out of alignment with it.

The cure for those fades would be to use a circularly polarized antenna. For HF, that's an item which can be created, at least on paper, but tends to be more than a little unwieldy at the scale of HF wavelengths. At the VHF and UHF frequencies we use for space propagation, though, circular polarization is far more practical and is practically mandatory.

To understand elliptical polarization, we must look a little deeper into the nature of radio waves. In a sense, there's no such thing as a linearly polarized radio wave. Any linearly polarized wave can be "decomposed" (physics word) into a right-hand circularly polarized wave and a left-hand circularly polarized wave. Down here on Earth that is normally irrelevant; everything balances out in the receiving antenna, before the signal even gets to the radio. In the ionosphere, though, the speed of propagation of those two waves is not equal. That's what really underlies Faraday rotation. Physicists say the waves have different "phase velocities." Because of this speed difference, their phase relationship varies as the waves pass through the ionosphere.

Some satellite communication systems actually turn this into an advantage by using the right-handed and left-handed waves to carry different information.

All of those effects except Faraday rotation can be caused by the entire atmosphere. However, the effects of the neutral atmosphere – air that isn't ionized – are negligible for our purposes. It's the ionosphere that creates the most disturbances to our space signal.

All these effects can significantly affect signals with frequencies as high as 12 GHz. They are especially significant when it comes to satellite links operating below 3 GHZ, and most of our satellite frequencies are *way* below 3 GHz.

Then there's Moon bounce. Adding the Moon into the equation obviously adds a tremendous amount of distance to the equation, but there are also some "Moon-specific" effects to consider if you're going to try Earth-Moon-Earth communications.

- Time spread. While most of the signal reflection takes place from a point in the center of the Moon's disc, the rest of the disc does provide some signal return. The edges of that disc are farther away from us than the center, to the tune of about 11 milliseconds of signal travel time (counting the outgoing and returning signal delays.)
For normal CW and other slower forms of modulation, this does not pose a practical problem. High data rates, though, may suffer from inter-symbol interference --- in plain language, your "a's" bleed over into your "b's and c's."

- Polarization changes. If you could get a perfectly linearly polarized signal up to the Moon and back, this wouldn't be a challenge, but Faraday rotation makes that unlikely. Circular polarization on the transmit and receive end can work well. There's a "however" on that, though. It goes like this: However, when a left-hand circularly polarized signal reflects off a surface, it becomes a right-hand circularly polarized signal. At the infinitesimal signal strengths of a returning Moon signal, that can become significant.

- Libration fading. We all know the Moon goes around the Earth about once a month, and that we always see the same side of the Moon. (That's because the Moon is what they call tidally locked to the Earth.) However, if you stuck a rod in the surface of the Moon pointing exactly at the center of the Earth, you'd find it wouldn't *quite* always point at the same spot. In fact, we don't just see 50% of the Moon. Over the course of a lunar month, we really can see very close to 59% of the Moon's surface. Why? The Moon, like everything else in orbit, isn't in a perfectly circular orbit, it's in an elliptical one, so it's constantly speeding up or slowing down relative to the Earth's rotation. It oscillates, just a little, relative to us. Like if you pointed your nose at a doorknob but slo-w-w-w-w-w-l-y moved your head just a little bit from side to side.

When we bounce signals off the Moon, we're bouncing them off a slowly moving target, and most likely we're bouncing off a spot that's an uneven surface. As that uneven surface moves relative to our antenna, the signal might exhibit a fluttery, irregular fading. Remember, we're working with very weak signals here, so it doesn't take much to upset our electromagnetic applecart.

Satellites

While we note all those potential foibles, at least one of our space propagation modes is reasonably simple to implement. The fact is satellites are just about as perfect a long distance propagation system as we're likely to see. We're not trying to get over a mountain, around a building, or through the ionosphere – we're not concerned about Fresnel zones, and the propagation really is close to 100% line of sight. That satellite sees a lot of territory!

The biggest challenges posed to the would-be satellite operator have to do with knowing when the satellites will be overhead and where they will be when they go over. After that it's "listen, point and shoot!" Work fast – it's "Call sign, grid square." That satellite is moving right along; most amateur satellites have orbital periods of around 90 minutes, so you only have a few minutes of access at a time.

Knowing the where and when of the satellite is easily available from AMSAT.org, various commercial software programs, or even apps for your smartphone like *Heavens Above*.

Working satellites requires neither Super Radio nor Super Antenna. It *does* usually require two radios, though, unless you happen to have a radio with full duplex capability; you'll need to send on the satellite's uplink frequency while monitoring on the downlink frequency.

While people have successfully made satellite contacts with a 5 watt handheld sporting a stock rubber ducky antenna, that takes just the right conditions and some luck. You'll most likely want a handheld Yagi – even a three element "tape measure" Yagi gives you a lot more gain than the average rubber ducky.

Overkill! A member of the Air Force Communications Command.

A friend of mine has been getting deeply into satellites lately. Here's his set-up. He has a homemade antenna that's two Yagis on one stick, for circular polarization,with elements for 2 meters and for 440. You can buy one from folks like Arrow Antennas, or find plans online to build your own.

The antenna sits on a camera tripod he got at the Goodwill store for $5.

Two $30 Baofeng handhelds hang by their hand straps off the tripod, and he has a little pocket tape recorder to catch the (very rapid) QSO's. He has a pair of headphones plugged into the receive unit, and hand holds the transmitter. That's it! Last time I talked to him he had worked something like 40 grid squares.

For a great "how to get started with satellites" guide, point your browser at:

http://www.work-sat.com/Home.html

Moon Bounce

While working satellites can be relatively easy and inexpensive, we can't honestly say the same about our other major form of space propagation, Earth-Moon-Earth, better known as EME or just moonbounce.

We can say, though, that it has gotten a whole lot easier and less expensive since the advent of the WSJT modes, and especially JT65B, designed specifically for moon bounce.

There are a number of challenges presented by EME, but by far the biggest is the loss created by the sheer distance the signal must travel. Your signal will lose 250 to 300 dB by the time it gets back to Earth. Compare that to a loss of around 150 to 160 dB for the longest hops here on Earth. It's not easy to make up 100 dB. To get some sense of the magnitude of the situation, the difference between a 5 watt QRP station and a 1500 watt boomer is a little shy of 25 dB.

The first moonbouncers used maximum legal power transmitters, and, often, arrays of high gain Yagi's or even big parabolic dishes. The only workable mode was CW. The advent of JT65 in 2003 dramatically changed things because it is even better in weak signal situations than CW – by about 15 dB.

That translates to us mere mortals being able to work moon bounce in favorable conditions with modestly powerful 2 meter (SSB) radios and amplifiers (150w to 500w), Yagis with gain of at least 10 dBd, a low noise amplifier (LNA) on the antenna, and a short-as-possible coax run.

According to the experts, you'll need a transmitter that will deliver at least 100 watts at the antenna – and keep in mind that JT65 has a 40% duty cycle in transmit mode, so be sure your transmitter/amp/antenna can handle that kind of load.

For a friendly step-by-step on making your first EME contacts, take a look at:

https://www.dxmaps.com/jt65bintro.html

Some of the versions of the widget you saw in the chapter on Space Weather also report EME predictions, and you'll find one on the home page of my web site here:

http://fasttrackham.com

Chapter 12 -- How we Figured out This Ionosphere Stuff

Early Hypotheses

The first person to publish a paper suggesting the existence of the ionosphere was Balfour Stewart, in 1882. Stewart was a Scottish physicist and meteorologist. He proposed that a conductive layer of upper atmosphere might account for the daily variations in the Earth's field that had been noted some 45 years earlier by Carl Friedrich Gauss.

Balfour Stewart

In 1889, Arthur Shuster, a British physicist, fleshed out Stewart's basic idea into a formal theory. He suggested the electric conductivity was a result of the sun's ultraviolet light and even proposed a formula for the variation of the conductivity which closely resembles modern theories of the form of the ionosphere.

Shuster never called it "the ionosphere" and wouldn't have been able to explain precisely how those ultraviolet rays somehow made air conductive. While Michael Faraday coined the term "ion" for the things-unknown that somehow carried electricity from one pole of a battery to the other, he had no more of a clue of what that term really meant in terms of atomic reality than anyone else on Earth at the time. It wasn't until 1884 that Swedish physicist Svante Arrhenius came up with a rough description of ions, but real understanding would have to await J.J. Thomson's discovery of the electron in 1894.

Arthur Schuster

In 1864, James Clerk Maxwell formulated his famous equations that predict the existence of electromagnetic waves, and in 1888, Heinrich Hertz proved they exist. Remember, Maxwell's equations say that radio waves and light waves are different frequencies of electromagnetic waves.

"Huh .... That's weird ....."

Imagine some folks' surprise when Marconi successfully sent radio waves *over* the horizon to a ship at sea. That shouldn't happen! You can't send a light beam from the shores of England over the horizon – there had to be some other phenomenon at play. Marconi thought radio waves would follow the curvature of the Earth, and some do, at least for a while, but his did not. (Yet another example of someone getting great results with a completely wrong theory.) Something up in the sky was bending those radio waves.

Back in 1888, Heinrich Hertz had done his famous experiments proving the existence of electromagnetic waves. In the course of those experiments, he also proved electromagnetic radiation in the microwave and radio regions of the spectrum displays the same basic behavior as visible light—reflection, refraction, diffraction, interference, polarization – and that they could be refracted by an electric field.

In 1902, in England, Oliver Heaviside, unaware of Schuster's theory, proposed that it was a conductive layer of the atmosphere that was bending the "Hertzian waves" back to Earth. At almost the same time, an American, Arthur Kennelly, also unaware of either Schuster or Heaviside, came up with the same idea, and thus for some time that theoretical layer was known as the Kennelly-Heaviside Layer, sometimes shortened to the Heaviside Layer.

Arthur Kennelly   Oliver Heaviside

In terms of scientific theory, that's where things sat for almost 25 years. Radio operators were gaining all sorts of practical experience in getting signals from one place to another, but not much of that accumulated wisdom made it into the scientific journals. Indeed, Marconi was notoriously guarded about sharing any technical details of his equipment at all; for instance, we still don't know for sure what frequency his 1901 trans-Atlantic signal used. (Truthfully, there's considerable doubt that he actually achieved that 1901 transmission at all.)

It was Sir Robert Watson-Watt who came up with the word "ionosphere" in 1926. Watson-Watt – a descendant of James Watt – would go on to be one of the chief inventors of radar, the achievement for which he was knighted.

Sir Robert Watson-Watt

Experimentalists

In 1924 English physicist Sir Edward Appleton, along with his graduate student Martin Barnette, began a series of experiments that would both prove the existence of the Kennelly-Heaviside Layer and begin to reveal its location and structure.

Sir Edward Appleton

Appleton's primary method was to transmit a continuous signal then measure the interference pattern created by the interaction of the ground wave of that signal with the wave reflected from up above.

Appleton started with a simple observation; a nearby BBC station could be heard clearly during the day but faded in and out at night. Appleton reasoned that the signal must be being conveyed to him along the ground during the day, but along two paths at night, and thus was going in and out of phase. He worked out a way to measure where that nighttime "extra" signal was coming from, and that's how he located what we now call the E layer of the ionosphere. Later he would find another layer above the E layer, now known generally as the F layer but also named the Appleton-Barnett layer.

Across the pond at the Carnegie Institution of Washington, Gregory Breit and Merle Tuve were also tracking down the ionosphere using a different technique. They used pulsed signals and timed how long it took the pulse to return.

Appleton would win the Nobel Prize for his efforts, and both his work and Breit-Tuve's were critical to Sir Robert Watson-Watt's breakthroughs in radar.

By 1932, Appleton's experiments had led him to write an extensive series of equations that predicted how radio waves would propagate via the ionosphere under varying conditions. Those equations are still foundational in the science.

Gregory Breit would go on to become, among other things, the first head of the Manhattan Project to create the atomic bomb. Merle Tuve founded the Johns Hopkins University Applied Physics Laboratory.

While scientists had long suspected that radiation hitting the upper atmosphere had *something* to do with the phenomena associated with the ionosphere, no one quite came up with a workable model until 1931 when English mathematician Sydney Chapman presented a theory that the ionosphere was formed by the action of ultraviolet light which ionized the atoms of air. He also came up with simple but surprisingly accurate mathematical models of the layering of the ionosphere.

It was Chapman, too, who first described some of the effects on the ionosphere of the Earth's magnetic field.

Watson-Watt, Appleton, and Chapman were all contemporaries and were well aware of and building on each other's work. Appleton's experiments, for instance, involved bouncing radio signals off a layer of the atmosphere then detecting and analyzing the return signal. It isn't a long leap from Appleton's signal bouncing to Watson-Watt's radar.

Sydney Chapman

By the middle of the 1930's, ionospheric observation stations had been improved and duplicated at numerous locations, so that we could map the ever-changing effects of the ionosphere in greater detail, and that greater detail showed the ionosphere to be far more complex and erratic than the simple, thin conductive layer envisioned by the early theorists.

Effects of geomagnetic storms had been mapped, as had sporadic E, and even Sudden Ionospheric Disturbances, which had been correlated to bright eruptions on the sun.

In 1936 a physicist at the U.S. Naval Research Laboratory, E.O. Hulburt, published a paper connecting variations in radio propagation via the ionosphere to the sunspot cycle. Hams have been hoping for sunspots ever since.

One key part of the puzzle that was still missing was a truly global map of the ionosphere, especially of the highest layer, the F2 layer. Because this map didn't exist, and because of the importance of the F2 layer to radio propagation, all this science failed to produce a body of practical advice for the would-be communicator. For the average radio operator, applied propagation science consisted mostly of, "Well, I dunno – this worked last time."

World War II & The Space Age

World War II stepped up the pace and the importance of ionospheric research. Allied and Axis power quickly built more and better ionosphere observation stations and more and better solar observatories. The main objective was to create reliable propagation forecasts, and to a large extent, the programs succeeded. Pure science took a back seat during the war, but practical application advanced considerably.

Of course, all this work was done from the ground, either by sending pulses or continuous waves skyward and attempting to deduce what mostly invisible forces were occurring some 100 miles above. One researcher reportedly said this was, "Like doing research by firing a gun in a dark room and waiting to hear who screamed."

Rocketry pioneer Robert Goddard had proposed studying the atmosphere with rockets as early as 1912, though his purposes were more along the lines of studying winds and pressures in the lower atmosphere. His ideas, for most of his life, generated little enthusiasm and less support, but World War II radically altered the US attitudes toward rocket science, and progress in the area was rapid. By 1946, a researcher at JPL, Frank Malina, had successfully headed up an effort to send the first scientific payload into near space. That particular rocket topped out at an altitude of 49 miles – not quite enough to get much information on the ionosphere – but the enormous possibilities were clear.

The USSR's successful launch of Sputnik I on October 4, 1957 put a supercharger on US space efforts, and by January of 1958 – after a spectacular launch failure in December 1957 – the US became the second nation to launch an artificial satellite. Of course, both the Soviet and US efforts had pretty much everything to do with military capability and pretty much nothing to do with pure science, but for political reasons both were announced as scientific projects and carried at least some instrumentation for scientific observations. Satellites proved to be extraordinarily useful. The very first US satellite, Explorer I, revealed the previously unknown Van Allen radiation belts.

The first satellite designed to study the ionosphere was the Canadian Alouette 1. Alouette launched in 1962. It carried equipment that, for the time, was very sophisticated, including two big dipoles (22.8 and 45.7 meters long) that deployed after the satellite reached orbit and were used to make measurements of the ionosphere from above. The main instrument on board was called a "topside sounder." It "pinged" the ionosphere on frequencies ranging from 1 MHz up to 12 MHz, and measured the return signal from above.

Alouette 1's mission lasted far beyond the most optimistic projections – it was still sending down data 10 years after launch. Until Alouette, the ionosphere had only been measured to a height of some 400 kilometers. The Alouette program extended that range to 3,500 kilometers, and vastly increased our understanding of many aspects of the ionosphere. (That extreme upper ionosphere isn't very directly relevant to radio propagation, though.)

Alouette 1 was the first of a fleet of satellites deployed for ionospheric studies, including the US Explorer 20 and Explorer 31. As I write this, a new ionospheric study satellite program, ICON, is being readied for launch, and should be in orbit by the time you read this. It will be used to study the interaction of the lower atmosphere with the ionosphere. Over the years it has become more and more apparent that solar activity doesn't quite account for everything that happens in the ionosphere and that terrestrial weather has some profound effects. However, the precise nature and mechanisms of those effects has eluded us due to lack of data.

When it launches, ICON will team up with another satellite already in orbit that also has a mission of studying the ionosphere. GOLD – Global-scale Observations of the Limb and Disk – gives us a "wide angle" picture of the ionosphere, while ICON will give us a close-up view. The combination of the two missions will give us, according to NASA, "the most comprehensive ionosphere observations we've ever had."

Incoherent Scatter Radar

As we were launching satellites, there was a breakthrough in ground-based observation of the ionosphere, as well, and some scientists claim it is the single most powerful instrument for those ground-based observations.

Around the time of the planning for Alouette, an American electrical engineer named William Gordon conceived of another, ground-based way to measure the ionosphere. His concept is called *incoherent scatter radar*.

William Gordon (Center) with engineer Domingo Albino

To understand incoherent scatter radar, let's start with a look at normal radar.

You probably know that radar works by sending out a radio signal then detecting the reflection of that signal off an object. When a radar signal hits a solid, relatively stable object, the return signal is a close match to the outgoing signal in terms of frequency and bandwidth. It might be slightly shifted by the Doppler effect if the object is in motion, but it still looks a lot like the original signal. Physicists would say it is *coherent*.

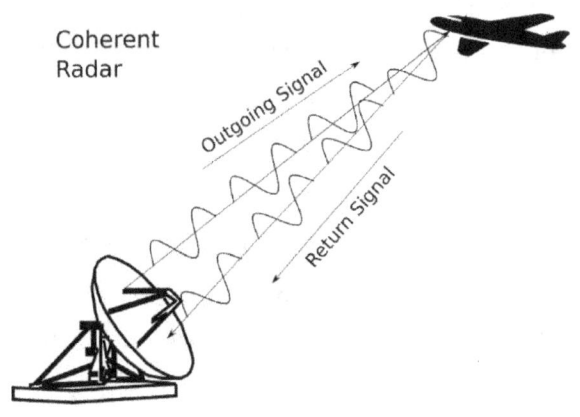

The ionosphere, though, is neither solid nor stable. In fact, it is neither solid, liquid, nor gas but a fourth state of matter called a plasma. In a plasma, electrons have been ripped from their atoms and are flying around in all sorts of different, though not random, directions.

With a sufficiently high power transmitter, a very high gain transmitting antenna, and most of all a very large receiving antenna, we can still bounce a radio wave off that plasma and detect the return signal. However, the return signal will be anything but coherent. Each electron that contributes to reflecting the return signal will add its own Doppler shift to that return signal. What comes back is an *incoherent* signal, consisting of a relatively broad band of different frequencies.

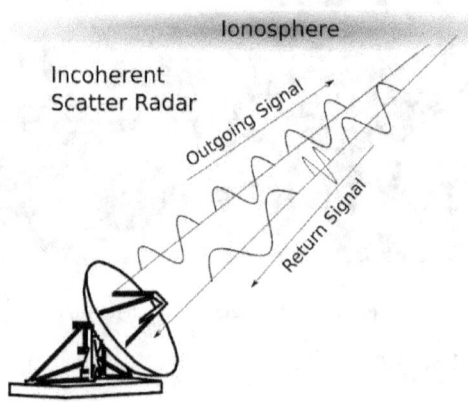

Bill Gordon's insight was to realize that the amount of incoherency in that return signal could be used to deduce several qualities of the point being measured. Of course, as with any radar, finding the altitude of the point being measured was fairly simple – since radio waves travel at the speed of light, it's just a matter of timing how long it takes to get a return signal.

By measuring the strength of the returned signal, one could know the *electron density* – one of our chief concerns, because it profoundly affects propagation. The bandwidth of the returned signals tells the *electron temperature*. While it seldom concerns us, knowing the temperature is important to researchers.

Later it would become clear that the incoherent scatter radar could also determine the direction of the wind currents in the ionosphere as well.

Bill Gordon quickly found himself on his way to Puerto Rico where he would oversee the construction and early years of operation of what was the world's first incoherent scatter radar and is now the Arecibo Observatory, a massive radio-telescope built in a 1000 foot wide sink hole in the Puerto Rican jungle.

Gordon went on to a distinguished career at Rice University and made many contributions to ionospheric science.

Today, massive incoherent scatter radar installations are scattered around the globe, covering all latitudes, feeding data to scientists who are still decoding the secrets of the complicated machine that is our ionosphere.

Arecibo Observatory

Afterword

We've covered a lot of territory! We've gone from below the ground to the moon and back, and we still haven't begun to cover everything there is to learn about propagation.

The truth is, you can get your EE degree with a specialty in propagation and you *still* won't have covered everything. More discoveries are being made all the time, too.

For us, as amateurs, there's a very practical side to learning all we can about propagation, but I think we also arrive at the hobby with a certain sense of wonder about the apparent magic of invisible waves flying through the air across town or across the globe. I know that fascination bit me very early – about as soon as I learned that there wasn't a little man inside that box that was playing music and talking to me! Since those days, there haven't been many days when I wasn't somehow involved in receiving and sending those magic waves. Strangely enough, even as the years have passed and I have learned more and more about how the magic works, that magic still amazes me as much as ever.

If nothing else, I hope this book has stoked that sense of wonder in you, and tickled your urge to learn more. I urge you to keep exploring, and remember; the only propagation method guaranteed not to work is the propagation method called "never turn on your transmitter."

Go forth and make some waves!

73,
Michael Burnette, AF7KB

Index

About the Author

Michael Burnette, AF7KB, started playing with radios at age 8 when he found the plans for a crystal radio set in a comic book and threw out a half roll of brand new toilet paper to get the cardboard tube for a coil form. He's been fascinated by radio ever since.

This beginning blossomed into a 25 year career in commercial broadcasting where he did a bit of everything from being an all-night DJ to serving as a vice president and general manager with Westinghouse Broadcasting (now CBS/Infinity.)

In 1992, Burnette left the commercial radio business behind, and took to traveling the world designing and delivering experiential learning seminars on leadership, management, communications, and building relationships.

He has trained people across the US and in Indonesia, Hong Kong, China, Taiwan, Mexico, Finland, Greece, Austria, Spain, Italy, and Russia. In addition to his public and corporate trainings, he has been a National Ski Patroller, a Certified Professional Ski Instructor, a Certified In-Line Skating Instructor, a big-rig driving instructor, and a Certified Firewalking Instructor.

He is the author of:

The Fast Track to Your Technician Class Ham Radio License
The Fast Track to Mastering Technician Class Ham Radio Math
The Fast Track to Your General Class Ham Radio License
The Fast Track to Mastering General Class Ham Radio Math
The Fast Track to Your Extra Class Ham Radio License
The Fast Track to Mastering Extra Class Ham Radio Math
The Fast Track to Understanding Ham Radio Propagation

These days he makes his home in the Seattle, WA area with his wife, Kerry (KG7NVJ) and their cat, who has no job, no radio, and no ham license.

He and Kerry make frequent appearances and presentations at major hamfests. They are also very active in their local ham club, where Michael works with new hams and Kerry serves as net control for a weekly YL net.

If you feel a need to communicate with Michael, his e-mail is AF7KB@fasttrackham.com.

See you at the next hamfest!

Credits

Map of geographic, magnetic and geomagnetic poles: By Cavit [CC BY 4.0 (https://creativecommons.org/licenses/by/4.0)], from Wikimedia Commons

Sunspot: By NASA [Public domain]

Solar Flare: By NASA [Public domain], via Wikimedia Commons

Magnetosphere: By NASA (http://sec.gsfc.nasa.gov/popscise.jpg) [Public domain], via Wikimedia Commons

Selected Bibliography

William Edwin Gordon January 8, 1918–February 16, 2010 BY Marshall. H. Cohen and Neal, F. L – National Academy of Sciences

World map: By Harbin (BlankMap-World-v2.png) [Public domain], via Wikimedia Commons

Magnetic, geomagnetic, and geographic pole positions: By Cavit [CC BY 4.0 (https://creativecommons.org/licenses/by/4.0)], from Wikimedia Commons

Magnetic fields attracting and repelling: Geek3 [GFDL (http://www.gnu.org/copyleft/fdl.html) or CC BY-SA 3.0 (https://creativecommons.org/licenses/by-sa/3.0)], from Wikimedia Commons

Understanding HF & VHF propagation conditions using data from N0NBH's HAMQSL Solar Data Panel chart ©N0NBH Paul L Herrman 2010 with author's permission.

All other images public domain or author's own work.

www.ingramcontent.com/pod-product-compliance
Lightning Source LLC
Chambersburg PA
CBHW071312220526
45468CB00001B/337